P9-CSW-653

3 –

50 genetics ideas you really need to know

Mark Henderson

Quercus

Contents

Introduction

We are living through a revolution in human understanding. For as long as our species has been capable of complex thought, people have wondered where we come from, why we behave as we do, how our bodies work in sickness and in health, and why we all seem so similar and yet display such diverse and wonderful individuality. Philosophy and psychology, biology, medicine and anthropology, even religion, have all attempted to supply answers, and not without some success. But until very recently, we have been missing a fundamental piece of the puzzle, with significance for every aspect of human existence. A knowledge of our genetic code.

Genetics is a young science. It is little more than 50 years since Crick and Watson discovered the "secret of life" – the structure of the DNA molecule in which organisms' cellular instructions are written. The first, incomplete, draft of the human genome was published only in 2001. Yet this infant branch of knowledge is already changing the way we understand life on Earth, and genetic technology is transforming the way we live.

Genetics has shed fresh light on our history, proving the fact of evolution, and allowing us to trace how the first humans emerged from Africa to populate the world. It has brought us new forensic tools that can convict the guilty and exonerate the innocent. And it is explaining how our individuality is forged through nature and nurture. We are also entering a new era of genetic medicine, which promises treatments tailored to patients' genetic profiles, spare part tissue grown from stem cells, gene therapy to correct harmful mutations, and tests that highlight inherited health risks that might then be reduced.

These incredible opportunities also stir ethical concerns. Issues such as genetic engineering, cloning, genetic discrimination and designer babies often suggest that DNA stands not for deoxyribonucleic acid, but for controversy.

We are, of course, much more than the sum of our genes. We are rapidly learning that other parts of the genome, such as the segments once derided as "junk DNA", are also important, perhaps more so. And as we grasp more about genetics, we are enhancing our understanding of other factors that count too – our lifestyles, our environments, our interactions with other people.

Without genetics, however, we would look at life with one eye. We are lucky enough to live at a time when humanity can finally watch with two.

01 The theory of evolution

Charles Darwin: 'There is grandeur in this view of life . . . that . . . from so simple a beginning endless forms most beautiful and most wonderful have been, and are being, evolved.'

'Nothing in biology', wrote the geneticist Theodosius Dobzhansky, 'makes sense except in the light of evolution.' It is a truth that applies particularly strongly to its author's specialist field. Though Charles Darwin had no concept of genes or chromosomes, those concepts and all the others that will be described in this book have their ultimate origins in the genius of his insights into life on Earth.

Darwin's theory of natural selection holds that while individual organisms inherit characteristics from their parents, they do so with small and unpredictable alterations. Those changes that promote survival and breeding will multiply through a population over time, whereas those that have negative effects will gradually disappear.

As is often the case with truly great ideas, evolution by natural selection has a beautiful simplicity that, once grasped, immediately becomes compelling. When the biologist Thomas Henry Huxley first heard the hypothesis presented, he remarked: 'How extremely stupid not to have thought of that!' Once a sceptic, he became evolution's most vociferous champion, earning the nickname 'Darwin's bulldog' (see box).

The argument from design For centuries before Darwin, natural philosophers had sought to explain the extraordinary variety of life on Earth. The traditional solution, of course, was supernatural: life, in all its diversity, was created by a god, and the traits that fit a particular organism to an ecological niche were a function of the creator's grand plan.

timeline

1802

William Paley (1743–1805) uses 'watchmaker analogy' to make the argument from design

Jean-Baptiste Lamarck (1744–1829) sets out theory of inheritance of acquired characteristics

1842

Charles Darwin (1809–82) sketches out evolution by natural selection in letter to Charles Lyell

This 'argument from design' dates back at least to the Roman orator Cicero, but it is most commonly associated with William Paley, an English clergyman. In an 1802 treatise, he likened the intricacy of life to a watch found on a heath, the very existence of which presupposes the existence of a watchmaker. It rapidly became scientific orthodoxy – even Darwin was much taken with it early in his career.

As was already clear to the philosopher David Hume in the 18th century, however, the argument from design begs the question: who designed the designer? The absence of an obvious naturalistic explanation for a phenomenon is a poor reason to look no further. Those who make it, from Paley to today's rebranded 'intelligent design' creationists, are essentially saying: 'I don't understand, so God must have done it.' As a way of thinking, it is no substitute for science.

Acquired characteristics While Paley was invoking the watchmaker, Jean-Baptiste Lamarck took a more intellectually curious approach to the problem. Organisms, he suggested, were descended one from another, with differences emerging by means of subtle modifications in each generation. His was the first theory of evolution.

Darwin's bulldog

T.H. Huxley won his nickname, 'Darwin's bulldog', during the 1860 meeting of the British Association for the Advancement of Science, when he defended Darwin's theory against the argument from design advocated by Samuel Wilberforce, the Bishop of Oxford. Though no verbatim account exists, Wilberforce began to mock his rival, asking whether he claimed descent from an ape through his mother or father. Huxley is said to have replied: 'I would rather be descended from an ape than from a cultivated man who used his gifts of culture and eloquence in the service of prejudice and falsehood.'

1858
Theory of natural selection presented to Royal Society by Darwin and Alfred Russel Wallace (1823–1913)

1859
Charles Darwin publishes *On the Origin of Species*

Lamarck's evolutionary driver was the inheritance of acquired characteristics: anatomical changes caused by the environment would be passed on to offspring. The son of a blacksmith would inherit the strong muscles his father built up in the forge. Giraffes stretch their necks to reach higher branches, elongating the necks of subsequent generations of calves.

The theory is often lampooned today, not least because of its revival in the 1930s by Stalin's favourite biologist, Trofim Lysenko. His insistence that wheat could be trained to resist cold snaps caused millions of deaths from famine in the Soviet Union. Lamarck's ideas are sometimes even described as heresy. Yet though he was wrong about the details of evolution, his broader thinking was astute. He correctly ascertained that biological characteristics are inherited – a perception of vast importance. He was mistaken only about the means.

On the Origin of Species The real means were soon elucidated by Darwin. In the early 1830s, he had sailed aboard the marine survey ship HMS *Beagle*, as naturalist and 'gentleman companion' to its captain Robert FitzRoy, on a voyage that enabled Darwin to make detailed observations of the flora and fauna of South America. He found particular inspiration in the Galapagos Islands, west of Ecuador, each of which was home to subtly different species of finch. Their similarities and differences led him to consider whether these species might be related, and had become adapted over time to the environment of each island.

In this, Darwin's assessment differed little from Lamarck's. What set his hypothesis apart was its mechanism. The economist Robert Malthus (1766–1834) had described how populations that increase in size would compete for resources, and Darwin now applied this principle to biology. Chance variations that helped an organism to compete for food and mates

Only a theory

Creationists like to dismiss evolution as 'only a theory', as if this gives their alternative scientific parity. This reflects their overwhelming misunderstanding of science, which does not use the term 'theory' in its common sense of a hunch. Rather, it means a hypothesis that is confirmed by all available data. Evolution more than meets this definition – it is supported by evidence from genetics, palaeontology, anatomy, zoology, botany, geology, embryology and many other fields. If the theory were wrong, almost everything we know about biology would have to be reassessed. It is like the theory of gravity – not an idea we can take or leave, but the best explanation currently available for an observed set of facts.

would help it to survive, and to pass those traits to its offspring. Variations with negative effects, however, would die out over time as their carriers lost out to others better adapted to their surroundings. Changes were not caused by the environment, but selected by it.

The implications of this natural selection were brutal. It worked towards no goal or purpose, and gave no special consideration to human life. What mattered, in Herbert Spencer's famous phrase, was 'the survival of the fittest'.

Darwin first sketched out his ideas as early as 1842, but he did not publish for another 17 years, fearing the derision that had been heaped upon treatises such as *Vestiges of the Natural History of Creation*, an 1844 pamphlet which argued that species can morph into new ones. In 1858, however, two years after starting to write up his theory, he received a letter from Alfred Russel Wallace, a younger naturalist who had developed similar notions. After presenting jointly with Wallace to the Linnean Society of London, Darwin rushed *On the Origin of Species* into print in 1859.

> **'The theory of evolution by cumulative natural selection is the only theory we know of that is in principle capable of explaining the existence of organised complexity.'**
>
> **Richard Dawkins**

Clerical naturalists, including Darwin's old tutors Adam Sedgwick and John Stevens Henslow, were outraged by the new theory. Another critic was Robert FitzRoy, who considered himself betrayed by an old friend, who had abused his kindness to promote views akin to atheism. But Darwin's theory found favour with a younger generation of intellectuals, who recognized both the theory's importance and its utility in undermining a scientific establishment that was still under heavy church influence.

The theory has been updated since 1859, not least by Darwin himself: in *The Descent of Man* (1871), he described how mating preferences can drive evolution as well as the environment, adding sexual selection to the scientific lexicon. But the central principle that all species are interrelated, and emerge one from another by way of random changes that are passed on if helpful to survival or breeding, has become the glue that holds biology together. It is also the foundation stone of genetics.

the condensed idea
Natural selection forms new species

02 The laws of inheritance

William Castle: 'What will doubtless rank as one of the great discoveries in biology, and in the study of heredity perhaps the greatest, was made by Gregor Mendel, an Austrian monk, in the garden of his cloister, some forty years ago.'

For all Charles Darwin's brilliance, his theory still lacked something critical at its core: it had no way of accounting for the individual variations that were supposed to be passed on from one generation to the next. Darwin himself favoured 'pangenesis', the idea that the characteristics of each parent merge in the offspring, but he was as wrong in this as was Lamarck about acquired characteristics. If only he had read a paper by a contemporary, a Moravian monk by the name of Gregor Mendel.

In 1856, the same year in which Darwin started work on *On the Origin of Species*, Mendel began a remarkable series of experiments in the garden of the Augustinian monastery of St Thomas in Brünn, now Brno in the Czech Republic. Over the next seven years, he was to breed more than 29,000 pea plants, with results that would make him known – when the rest of the world finally took notice – as the founder of modern genetics.

Mendel's experiments Botanists had long known that certain plants 'breed true' – that is, characteristics such as height or colour are reliably transmitted to the next generation. Mendel exploited this in his experiments on variation, by taking seven true-breeding pea traits, or phenotypes, and cross-breeding the plants that bore them to create hybrids. Pea strains that always produced round seeds, for example, were crossed with strains with wrinkled seeds; purple flowers with white; tall

1856
Gregor Mendel (1822–84) begins breeding experiments with pea plants

1865
Mendel presents laws of inheritance to Natural History Society of Brünn

Mendelian Inheritance in Man

The Online Mendelian Inheritance in Man database includes more than 12,000 human genes that are thought to be passed on according to Mendel's laws, with dominant and recessive alleles. Of these, at the time of writing 387 variable genes have been sequenced and linked to a specific phenotype, including diseases such as Tay–Sachs or Huntington's, and more neutral traits such as eye colour. Several thousand other phenotypes are known to follow a Mendelian inheritance pattern, but the parts of the genome responsible have yet to be identified or mapped. About 1 per cent of births are affected by Mendelian disorders, which result from variation in a single gene.

stems with short. In the next generation, known to geneticists as F_1, only one of the traits would remain – the progeny always had round seeds, purple flowers or tall stems. Parental characteristics did not blend, as pangenesis suggested, but one characteristic invariably seemed to dominate.

Next, Mendel took each hybrid and used it to fertilize itself. In this F_2 generation, the trait that seemed to have been erased suddenly came back. Around 75 per cent of the peas had round seeds, with the remaining 25 per cent coming out wrinkled. In all seven of his samples, this same ratio of 3:1 emerged. His results fit the pattern so well, indeed, that some later scientists have suspected fraud. The principles he discovered are now too well attested for that, but it is quite possible that Mendel realized the implications of this ratio early on, and stopped experimenting just when the numbers added up nicely.

First experiment

Round pea Wrinkled pea

Two homozygous dominant alleles (each denoted as **R**)

Two homozygous recessive alleles (each denoted as **r**)

Round peas

In the F_1 generation, all offspring are heterozygous, with one allele of each kind. The peas are round as the round allele, **R**, is dominant

Second experiment: uses offspring of first experiment

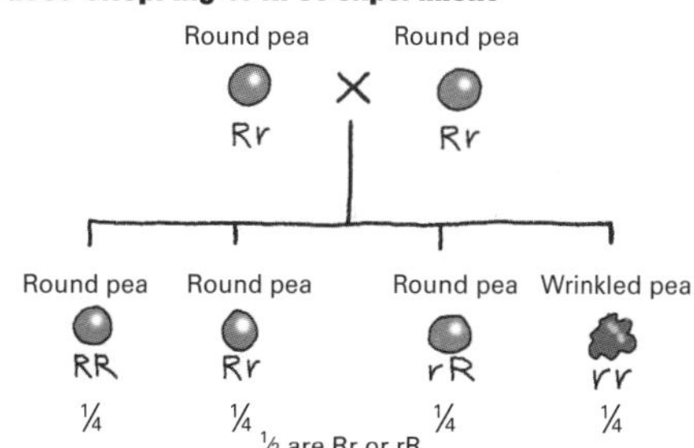

In the F_2 generation, the ratio of round (dominant) to wrinkled (recessive) peas is 3:1

1900

Rediscovery of Mendel's ideas by Hugo de Vries, Carl Correns and Erich von Tschermak

What was happening, Mendel recognized, was that these phenotypes were being transmitted by paired 'factors' – we would now call them genes – some of which are dominant and others recessive. The parent plants bred true because they carried two dominant genes for round seeds or two recessive genes for wrinkled – in the language of genetics, they are homozygous. When they were crossed, the progeny were heterozygous – they inherited one gene of each kind. The dominant gene won out, and all the seeds were round.

In the F_2 generation, there were three possibilities. A quarter, on average, would have two round-seed genes, and thus round seeds. Half would have one gene of each sort, producing round seeds because that gene was dominant. And another quarter would inherit two wrinkled-seed genes, producing wrinkled seeds: such recessive genes can generate a phenotype only when no dominant gene is present.

Mendel's laws Mendel used these results to draw up two general laws of inheritance (to avoid confusion, I will use the language of modern genetics here and not his own). His first principle, the law of segregation, holds that genes come in alternative varieties known as alleles, which influence phenotypes such as seed shape (or eye colour in people). Two alleles governing each phenotypic trait are inherited, one from each parent. If different alleles are inherited, one is dominant and expressed, while the other is recessive and silent.

complex dominance

Not all traits that are governed by single genes follow quite the inheritance pattern that Mendel discovered. Some genes are incompletely dominant, meaning that when an organism is heterozygous, with one copy of each allele, the resulting phenotype is intermediate. Carnations with two alleles encoding red colour are red; those with two white alleles are white; and those with one of each are pink. Genes can also be co-dominant, meaning that heterozygotes express both traits. Human blood groups work like this: while the O allele is recessive, the A and B alleles are co-dominant. So both the A and B alleles are dominant to O, but a person who inherits one A and one B will have type AB blood.

His second principle was the law of independent assortment: the inheritance pattern of one trait does not influence the inheritance pattern of another. The genes that encode seed shape, for example, are separate from the ones that encode seed colour and will not affect it. Each Mendelian trait will be passed on in 3:1 ratios, according to the dominance pattern of the genes involved.

Neither of Mendel's laws is quite correct. Some phenotypes are linked and often inherited together – as are blue eyes and blond hair among Icelanders – and not all traits follow the simple patterns of dominance found in the monk's peas. But they were a good first effort. Genes found on different chromosomes are indeed inherited separately in line with the second law, and there are plenty of diseases that fit with the first. These are known as Mendelian disorders – conditions such as Huntington's disease, which always occurs in people who have one copy of a dominant mutated gene; or cystic fibrosis, caused by a recessive mutation that is dangerous only when two copies are inherited, one from each parent.

‘Mendelism supplied the missing parts of the structure erected by Darwin.’

Ronald Fisher

Rejection, ignorance and rediscovery Mendel read his paper on inheritance to the Natural History Society of Brünn in 1865, and it was published the following year. But if Darwin's opus was a sensation, Mendel's was barely read, and those who did read it missed its significance. It appeared, indeed, in a volume in which Darwin annotated both the preceding and following articles, but he left the work that was ultimately to underpin his theory unmarked. In 1868, Mendel was elected abbot and his research ceased, though he remained certain of its significance. Shortly before his death in 1884 he is said to have remarked: ‘My scientific work has brought me a great deal of satisfaction, and I am convinced that it will be appreciated before long by the whole world.’

He was right. In the 20th century, Hugo de Vries, Carl Correns and Erich von Tschermak each developed similar theories of inheritance to Mendel's, and acknowledged the monk's priority. A new science was born.

the condensed idea
Genes can be dominant or recessive

03 Genes and chromosomes

C.H. Waddington: 'Morgan's theory of the chromosome represents a great leap of imagination comparable with Galileo or Newton.'

When T.H. Morgan (1866–1945) began experimenting with fruit flies in 1908, he accepted neither Darwin nor Mendel. Although he was satisfied that some form of biological evolution had taken place, he doubted natural selection and Mendelian heredity as the means. His results, however, were to convince him that both theories were in fact correct, and revealed the cellular architecture that allows traits to be conveyed from one generation to the next.

Morgan proved not only that phenotypes are inherited in the fashion that Mendel proposed, but also that the units of heredity reside on chromosomes. These structures in the cell's nucleus, of which humans have 23 pairs, had first been discovered in the 1840s, but their function remained uncertain. In 1902, biologist Theodor Boveri and geneticist Walter Sutton independently proposed that chromosomes might hold the material of inheritance, to great controversy. Though Morgan was among the sceptics, his fruit flies put the argument to rest. He provided the physical evidence that cemented the Mendelian revolution.

The field of study that it threw open now had a name. Mendel had called the codes for heritable traits 'factors', but in 1889, before his role in rediscovering the monk's work, Hugo de Vries had used 'pangen' to describe 'the smallest particle [representing] one hereditary characteristic'. In 1909, Wilhelm Johannsen produced a more elegant contraction – the gene – along with the term 'genotype' to signify an organism's genetic

timeline

1840s
Discovery of chromosomes

makeup, and 'phenotype' to signify the features that genes produce. William Bateson, an English biologist, rolled all this together into a new science: genetics.

The threads of life Chromosomes, we now know, are threads composed of chromatin – a combination of DNA and protein – that sit in the cell's nucleus and hold the vast majority of its genetic information (a little lies elsewhere, in mitochondria and chloroplasts). They are usually depicted as sticks that are pinched in the middle, but they actually take on this form only during cell division. For most of the time, they are long, loose strings, like fabric necklaces. Genes are like patches of colour woven into the design.

The number of chromosomes differs from organism to organism, and they almost invariably come in pairs: individuals inherit one copy from their mother and one from their father. Only in reproductive cells called gametes – in animals, the eggs and sperm – is just a single set present. Ordinary paired chromosomes are known as autosomes, of which humans have 22 pairs, and most animals also have sex chromosomes that can differ between males and females. In humans, people who inherit two X chromosomes are female, while those who have one X and one Y are male.

Chromosomal disorders

Inherited diseases are not always caused by mutations in specific genes; they can also be caused by chromosomal abnormalities, or aneuploidies. One example is Down's syndrome, caused when people inherit three copies of chromosome 21 instead of the usual two. This extra chromosome causes learning disabilities, a characteristic physical appearance and an increased risk of heart disorders and early-onset dementia.

Aneuploidies of other chromosomes are almost invariably fatal before birth. They are often responsible for miscarriage and infertility, and it is becoming possible to screen in-vitro fertilization (IVF) embryos for these faults to improve couples' chances of a successful pregnancy.

1902

Theodor Boveri (1862–1915) and William Sutton (1877–1916) suggest chromosomes could carry genetic material

1910

T.H. Morgan (1866–1945) proves chromosomal basis of inheritance

Humans and other animals

Humans have 23 pairs of chromosomes – the 22 autosomes, plus the sex chromosomes X and Y. Until 1955, however, it was widely agreed that we had 24 pairs, in common with our closest animal relatives, the chimpanzees and other great apes. This was overturned when Albert Levan and Joe-Hin Tjio used new microscopy techniques to reveal 23 pairs. Closer examination of the human chromosome 2 shows that it was formed by the fusion of two smaller chromosomes that still exist in chimpanzees. This merger was one of the evolutionary events that made us human.

In the 1880s, the advent of dyes that can stain chromatin allowed the embryologist and cytologist Edouard van Beneden to observe that each cell's maternal and paternal chromosomes remain separate throughout cell division, a discovery that led Boveri and Sutton to suggest a role in Mendelian inheritance. If genes were held on discrete chromosomes that came from each parent, that could explain how recessive traits could be preserved to re-emerge in later generations.

The fly Boveri and Sutton were proved right by one of their biggest critics – Morgan. His instrument was the humble fruit fly, *Drosophila melanogaster* – the Latin name means 'black-bellied dew-lover'. Females can lay 800 eggs a day, and its fast reproductive cycle, which can produce a new generation every two weeks, allowed Morgan's lab to cross-breed millions of the insects to examine patterns of inheritance.

Drosophila usually has red eyes, but in 1910 Morgan found a single white-eyed male. When he mated the mutant with an ordinary red-eyed female, their progeny (the F_1 generation) were all red-eyed. These flies were then mated with one another, to produce the F_2 generation in which Mendel's recessive traits reappeared. The white-eyed phenotype came back – but only in about half of the males, and in none of the females. This result seemed to be linked to sex.

'Morgan's findings about genes and their location on chromosomes helped transform biology into an experimental science.'

Eric Kandel

In humans, sex is determined by the X and Y chromosomes – females are XX and males XY. Eggs always contain an X, while sperm can bear an X or a Y. As the X chromosome affects the sex of fruit flies in similar fashion, Morgan realized that his results could be explained if the mutant gene that turned eyes white was recessive, and was carried on the X chromosome.

In the F_1 generation, all the flies were red-eyed as they inherited an X chromosome from a red-eyed female, and thus had a dominant red-eyed gene. The females were all carriers of the recessive gene, but it was not expressed. None of the males had it at all.

In the F_2 generation, all the females were red-eyed as they received an X chromosome with a dominant gene from a red-eyed father – even if their mothers were carriers and passed on a mutant X, they would not develop white eyes as the trait is recessive. Among the F_2 males, however, the half that received a mutant X from their mothers were white-eyed: they had no second X chromosome to cancel out the effects of the recessive gene.

Morgan had hit on a critical principle. Many human diseases, such as haemophilia and Duchenne muscular dystrophy, follow this sex-linked pattern of inheritance: the rogue genes responsible lie on the X chromosome, hence the diseases are developed almost exclusively by males.

Genetic linkage As Morgan studied *Drosophila* still further, his team found dozens more traits that seemed to be carried on chromosomes. Sex-linked mutations were the simplest to spot, but it soon became possible to map genes to the autosomes, too. Genes that lie on the same chromosome tend to be inherited together. By studying how often certain fly traits are co-inherited, Morgan's 'drosophilists' were able to show that certain genes lie on the same chromosome, and even to calculate their relative distance from one another. The closer the genes, the more likely they are to be passed on together. This concept, called genetic linkage, is still a key tool for finding genes that cause disease.

Morgan had been wrong about Mendel, wrong about Boveri and Sutton, and indeed wrong about Darwin. But he was not pig-headed in his wrongness. Instead, he used experimental data to overcome it and develop a fundamental idea. His conversion is a perfect illustration of one of science's great strengths. Unlike in politics, when the facts change in science it's OK to change your mind.

First experiment

Red-eyed female × White-eyed male

(XX) RR

(XY) r–

Red-eyed females have two dominant red-eyed alleles, R, carried on their two X chromosomes

White-eyed males have one recessive white-eyed allele, r, on their X chromosome, and no eye colour gene on their Y chromosome (denoted –)

Red-eyed females

Red-eyed males

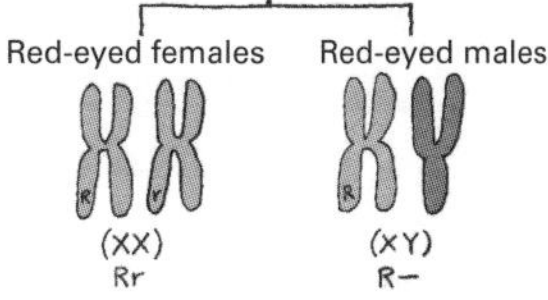

(XX) Rr

(XY) R–

In the F_1 generation, all flies are red-eyed, as they all have one copy of the dominant red-eyed allele, R

Second experiment: uses offspring of first experiment

Red-eyed female (XX) Rr × Red-eyed male (XY) R–

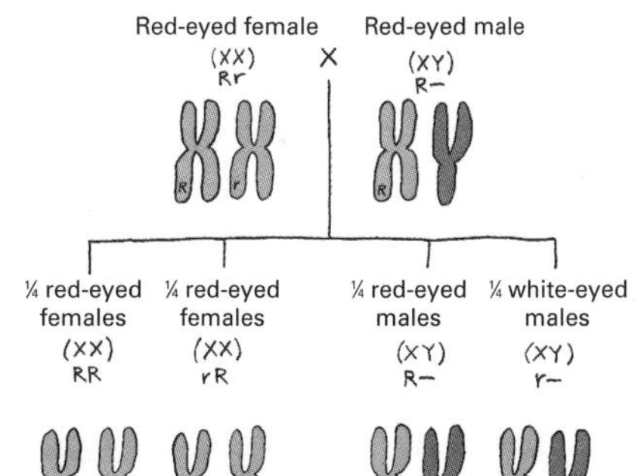

¼ red-eyed females (XX) RR

¼ red-eyed females (XX) rR

¼ red-eyed males (XY) R–

¼ white-eyed males (XY) r–

In the F_2 generation, all females are red-eyed as they have at least one dominant X-linked red-eyed allele, R. Half the males have the dominant allele, R, and are red-eyed, but half have the recessive allele, r, and are white-eyed

the condensed idea
Genes are carried on chromosomes

04 The genetics of evolution

Ernst Mayr: 'New gene pools are generated in every generation, and evolution takes place because the successful individuals produced by these gene pools give rise to the next generation.'

Mendelian genetics is today accepted as the mechanism by which Darwinian evolution occurs. Yet at the time of its rediscovery, Mendel's theory was widely considered incompatible with Darwin's. Attempts to reconcile the two great ideas of 19th-century biology were to become a dominant theme of early 20th-century genetics, sketching out principles that remain accepted in their essentials today. The resolution is known as the modern evolutionary synthesis.

Many of the biologists who first championed Mendel's ideas thought that his portrait of discrete genes seemed to rule out the gradual evolution proposed by natural selection. Mendelian inheritance did not appear to generate enough reliably heritable variations for slow selective processes to generate new species. Instead, the 'mutationists' or 'saltationists' suggested that sudden major mutations might move evolution forward in jumps.

A rival school, the biometricians, agreed with Darwin that there was wide and continuous variation between individuals, but drew from this that Mendel must be mistaken. Inherited traits, they felt, could not explain such variety if genetic information was carried by self-contained units, which could emerge intact after being hidden for a generation. There appeared to be too many differences between organisms within the same species, let alone between separate ones, for discrete genes to explain them all.

timeline

1859

Darwin publishes *On the Origin of Species*

1865

Mendel identifies laws of inheritance

The X-Men

The superheroes of the X-Men comics and films are supposed to have acquired their remarkable powers, such as Magneto's control over magnetic fields and Storm's ability to change the weather, through spontaneous genetic mutations. This is entertaining stuff but scientific nonsense – and not just because these powers are implausible. The narrative follows the heresy of saltationism – the notion that evolution takes sudden leaps forward when individuals acquire massive mutations that allow them to do new things. Population genetics killed this misconception in the early 20th century: evolution actually happens by way of slight mutations, which can cause rapid change as they are selected by the environment.

T.H. Morgan's discoveries about chromosomes started to explain how Darwin and Mendel might work together. His flies showed that mutations do not generate species by themselves, but rather increase the diversity of a population, providing a pool of individuals with different genes on which natural selection can act. This led a new generation of geneticists to realize that the two theories might be successfully combined. For the evidence, they turned to new tools, mastering mathematics and taking their research out into the field.

Population genetics The key to understanding how natural selection could work by Mendelian means was to move beyond the level of individual organisms and genes. This required two important insights. First, the English geneticist Ronald Fisher grasped that most phenotypic traits are not governed by a single gene in the neat fashion of Mendel's peas, but are instead influenced by a combination of different genes. He used new statistical methods to prove that this sort of inheritance could account for the wide variation between individuals measured by the biometricians, without invalidating Mendel's laws.

1910
Morgan's chromosome experiments suggest the two theories are compatible

1924
J.B.S. Haldane (1892–1964) publishes work on peppered moths

1930
Ronald Fisher (1890–1962) publishes *The Genetical Theory of Natural Selection*

1942
Julian Huxley (1887–1975) publishes *Evolution: The Modern Synthesis*

The population geneticists also realized that the emergence of mutations that produce new genetic variants or alleles is only the start of the evolutionary process. What matters more is how these alleles become distributed through entire populations. Very large mutations, of the sort considered crucial by the saltationists, are unlikely to spread: when not in themselves lethal, they tend to be so significant that they create individuals that are out of step with their environments. Such variants are less likely to survive and breed. Slight mutations that turn out to be advantageous, however, will gradually take over the gene pool, as those who carry them have more descendents.

The peppered moth The most celebrated example is the peppered moth. Before the Industrial Revolution in England, these insects were uniformly white-bodied and mottled, an adaptive colour scheme that camouflaged them against the lichen that covered tree trunks. Over the course of the 19th century, however, pollution from the mills of Manchester and other industrial centres blackened the local woods with soot and killed off lichen.

The peppered moth has a variant that is dark-coloured, owing to a mutation in the gene that produces the pigment melanin. These were very rare in the early 19th century, accounting for an estimated 0.01 per cent of the population: it was a prime example of a major mutation that reduced fitness, as black moths stood out and were quickly eaten by birds. By 1848, however, 2 per cent of moths in the Manchester region were black, and by 1895 the figure stood at 95 per cent. The changed environment, in which soot-covered trees now predominated, had given the dark allele an adaptive advantage.

The English geneticist J.B.S. Haldane calculated that the dark allele's almost total takeover of the moth population required black insects to be just 1.5 times more likely to survive and breed because of their colour. Mathematics has since shown that very small genetic changes of this sort can mushroom in frequency remarkably quickly, even if they have only slightly adaptive effects. Natural selection is a powerful engine, fuelled by genetics.

Genetic drift Natural selection is not the only method by which evolution occurs. Genes can also drift. As Mendel's law of segregation says, individuals have two copies of each gene, and pass one on to their offspring at random. In a large population, each allele will be passed on to subsequent generations in the frequency at which it was initially present, provided no selective pressures are at work. The randomness of this process, though, means that strange things can happen when populations are small. Chance variations in inheritance can cause one genetic variant to become more common than another, without any natural selection at all.

Imagine a bird species has two alleles for beak length, one long and one short, and that all the parents in a colony have one copy of each. In a big population, each allele will have a frequency of roughly 50 per cent in the next generation, because of the large numbers involved. Now imagine there are just two breeding pairs, again with one copy of each allele. The most likely outcome is still a 50–50 split, but the small numbers can't guarantee it. One allele might predominate in the offspring for no reason other than chance. Biologists call this the 'founder effect' – the gene pool of any new colony is shaped by the random genotypes that happen to be carried by its founders.

Speciation

One of the triumphs of the evolutionary synthesis was to provide understanding of how new species are formed. There are four main mechanisms by which this can take place, but they all rely on the partial or complete separation of two population groups, often by a geographical barrier such as a river or mountain range, so that they can no longer interbreed. Once these groups become isolated, genetic drift means they will become less and less similar over time, even if no selective pressures are involved. When these populations come back into contact with one another, they have often diverged so much that they cannot interbreed – they have turned into separate species.

This concept of genetic drift added another means by which Mendelian inheritance could account for variation within and between species, without recourse to sudden mutational leaps. Even in cases where natural selection did not appear to be happening, science had a second way in which evolution could be explained by genetics. The evidence that Mendelism and Darwinism were compatible was starting to stack up.

the condensed idea
Genetics drives evolution

05 Mutation

Hermann Muller: 'Mutation is indeed capable of being influenced "artificially" . . . it does not stand as an unreachable God playing its pranks upon us from some impregnable citadel in the germ plasm.'

The modern synthesis showed that large mutations are not the transforming force of evolution. Yet without genetic alterations of some sort, evolution could not happen. Natural selection and drift might be the processes that cause particular alleles to proliferate, but these alleles must somehow differ from other variants in the first place. The genetic code has to be copied faithfully from generation to generation for characteristics to be reliably inherited, but not too faithfully. The minor copying errors – small mutations – provide the raw material of evolution, the sparks that light its fire. They are then fanned into a blaze by natural selection; they slowly smoulder by drift; or they fail to catch light and die out.

T.H. Morgan's experiments with *Drosophila* struck gold because of a random mutation: the fly with white eyes. His team had increased the possibility of spotting such a chance event by breeding millions upon millions of the insects, but spontaneous mutations are so rare that vast numbers of specimens were needed to find them. Relying on chance, and time, made research exceptionally laborious. A means of kick-starting evolution by inducing mutations, however, was soon to transform the power of *Drosophila* research.

The breakthrough came from one of Morgan's students – a Jewish New Yorker named Hermann Muller. A brilliant theorist, Muller's ideas had proved vital to explaining the drosophilists' chromosomal work, but as he had not conducted the experiments himself he received little credit in the group's publications. Upset by his treatment – he was a cantankerous and

timeline

1910–15

Morgan shows chromosomal basis of inheritance

difficult man to work with, though biographers have also suggested he was a victim of anti-Semitism – Muller fell out with his teacher and moved to Texas to start working on his own.

X-rays Muller was fascinated by mutation, and also by Ernest Rutherford's recent achievement of splitting the atom. Like atoms, genes had been widely considered irreducible and immutable. If atomic form could actually be changed, might it be possible, he wondered, to alter genes artificially too? And might radiation be a potential agent of that change? In 1923, he began to expose fruit flies to radium and X-rays, to test his hypothesis.

The first experiments yielded little. While the X-rays seemed to be causing mutations, proof was elusive because they also had the unfortunate side-effect of sterilizing the insects, making it impossible to study what happened to their offspring. In November 1926, however, Muller finally got his radiation doses right. When he exposed male flies to X-rays and then mated them with virgin females, they bore mutant offspring at unprecedented frequencies. Within a few weeks, he had generated more than 100 mutants – half as many as all the spontaneous ones that had been identified in the previous 15 years.

Muller and Stalin

An ardent communist, Muller moved to work in the Soviet Union in 1935, where he developed a socialist approach to eugenics. Selective breeding, he argued, might be used in social engineering, to produce a new generation more inclined to live in accordance with the teachings of Marx and Lenin. Stalin, however, was unimpressed: under the influence of Trofim Lysenko, he declared genetics based on Mendel and Darwin to be bourgeois science, and began persecuting its practitioners.

Muller's collaborator, Nikolai Vavilov, was arrested and later died in the gulag (Soviet labour camp). Muller fled before he met with the same fate.

1927
Hermann Muller (1890–1967) shows that X-rays can induce mutation

1943
Max Delbrück (1906–81) and Salvador Luria (1912–91) show mutation is independent of natural selection

Some of the mutations were lethal, but many were not, and these were passed on reliably to later generations, as Mendel had predicted. Muller noted breakages in the flies' chromosomes, and his correct interpretation was that as radiation strikes them, it causes random alterations in their genetic structure.

Often, the changes that result are so damaging that they cause immediate death, or they are so maladaptive that they rapidly disappear from the gene pool. But occasionally, the outcome is a small 'point mutation' in an individual gene, leading to slight phenotypic variation of the sort that can spread through a population by natural selection or drift. Radiation can do this artificially, and quickly, in the laboratory. In the wild, the same thing is accomplished by random copying errors, or by exposure to environmental mutagens such as ultraviolet light or certain chemicals.

Genetic manipulation Muller immediately grasped the significance of his discovery. Science now had a tool for causing a mass of mutations in laboratory organisms, which greatly improved the speed and efficiency with which genetics could be studied. It was now a properly researchable subject. But the advance also suggested that if mutations could be induced, they could be manipulated too.

It meant that evolution could be artificially accelerated, by exposing organisms to radiation, and then breeding selectively from any mutants that had acquired desirable traits. When Muller spelled this out in a series of public speeches in the late 1920s, he became the first celebrity of genetics. He was the first to imagine the potential of genetic modification, long before the first genetically modified (GM) crop was ripped up by Greenpeace.

Muller proposed that radiation might be used to produce new agricultural varieties; other scientists soon demonstrated that it creates heritable mutations in maize. X-ray mutagenesis is still used today in just that fashion, to create new varieties for plant breeders (despite their unnatural origins, such crops are perfectly acceptable to organic farmers, though other approaches to genetic engineering, oddly, are not). Muller suggested that other applications might be found in medicine and industry, as indeed they were. He even thought that artificial mutations could be used to direct human evolution in positive ways.

The Luria-Delbrück experiment

While the importance of mutation to evolution was well established by the 1940s, a question remained. Did natural selection simply preserve random mutations that happened to be advantageous, or might selective pressures make mutation more likely to occur? Salvador Luria and Max Delbrück settled this in 1943, by experimenting with bacteria, and viruses that prey on bacteria known as phages. They found that mutations that made bacteria resistant to phages occurred randomly, and at a reasonably consistent rate, regardless of selective pressure. Mutation happens independently of natural selection, not because of it.

The dangers of radiation This last idea, however, would require a less hazardous means of inducing mutations than X-rays. For another implication of Muller's discoveries was that radiation is not normally a benign or neutral influence on genes. Most of the mutations it triggers in DNA (see Chapter 7) are not harmless or neutral, but catastrophic: large numbers of Muller's mutant flies died, and others were sterile. In longer-lived organisms than *Drosophila*, including humans, this sort of genetic damage commonly causes cancer. Muller began to campaign for greater public awareness of the risks of exposure to radiation, for example to doctors administering medical X-rays.

Geneticists, indeed, were to prove pivotal in building an appreciation of the dangers of radiation, particularly in the atomic era that followed the Manhattan Project of the Second World War and the bombing of Hiroshima and Nagasaki. Figures such as Muller and US scientist Linus Pauling used their understanding of the serious and irreversible damage that radioactivity can inflict on DNA in a successful campaign against atmospheric nuclear tests. Pauling won his second Nobel Prize, for peace, for his role. The benefits of Muller's X-ray experiments were not restricted to advances in genetics and plant breeding. They also made humanity aware of a grave threat to health.

the condensed idea
Mutations can be induced

06 Sex

Graham Bell: 'Sex is the queen of problems in evolutionary biology. Perhaps no other natural phenomenon has aroused so much interest; certainly none has sowed as much confusion.'

Sex is one of the great problems of life. That isn't just because of the amount of time we spend thinking about it, but also because it is an evolutionary and genetic puzzle.

Many organisms – most, indeed, given that bacteria make up a high proportion of the world's biomass – are perfectly happy procreating by themselves. Why, then, does asexual reproduction not apply across the board? It is good enough for most of the cells in the human body – the somatic cells that make up organs like the liver and kidneys divide as if they were asexual microorganisms. The only exceptions are our germ cells, which make the sperm and eggs (gametes) that ultimately make new humans.

Asexual reproduction allows any organism to duplicate its entire genome into its offspring, give or take a few random copying errors. Sex, however, means only half a population can bear young, reducing the reproduction rate, and it requires both sexes to spend time and energy finding partners. Only half a parent's genes get into its sons or daughters. All these things should be bad in terms of natural selection. Yet sex not only persists, it thrives – it is the reproductive system used by most visible life.

Its survival against the apparent odds is explained by what happens at a genetic level, and by what that means for evolution. Random mutations are not the only fuel for natural selection and drift. Sex causes variation too, by shuffling the genetic pack every time it happens. This process, known as crossing-over or recombination, repeatedly throws the code of life together in new arrangements, which can be passed on to future

timeline

1910
Morgan shows chromosomal basis of inheritance

1913
Morgan and Alfred Sturtevant (1891–1970) identify crossing-over and produce first genetic map

generations. Any that prove particularly advantageous will be favoured, just like beneficial mutations.

Meiosis and mitosis This opportunity for variation emerges from a special method of cell division that is unique to sex. The overwhelming majority of the human body's cells are diploid, with a full count of 46 chromosomes, arranged in two sets of 23 pairs. When these somatic cells divide, as the body grows or heals, they copy their genomes completely by a process called mitosis. All the pairs of chromosomes are duplicated, and the two sets are drawn apart as the cell splits, so that one set ends up in each of its daughters. The result is two new diploid cells, each with 46 chromosomes identical to those of their parent.

Mitosis, in its essentials, is asexual reproduction. And the one place in the body where it doesn't apply is the parts that are specialized for sex. In germ cells, eggs and sperm are made by another method of cell division – meiosis. During meiosis, the diploid precursor cells of gametes duplicate their DNA, then share it out equally between four daughter cells with 23 chromosomes each. In men, these become sperm, and in women, one becomes an egg while the other three are discarded as 'polar bodies'.

Cells of this type are known as haploid – they have just one copy of each chromosome, instead of the pairs found in diploid somatic cells. When the two types of gamete fuse after sex to generate an embryo, the full complement of 46 is restored, with one copy of each chromosome furnished by each parent.

Crossing-over This fusion of genetic material from two individuals provides variation, by creating different combinations of chromosomes. But it is not the only way in which sex adds value: the actual composition of the chromosomes that go into each sperm and egg is also unique.

When paired chromosomes line up during meiosis, they swap genetic material between themselves. The two strands of DNA – one originally inherited from an individual's mother and one from its father –

Recombination

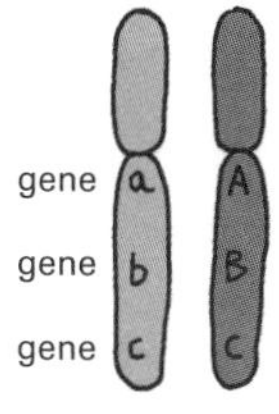

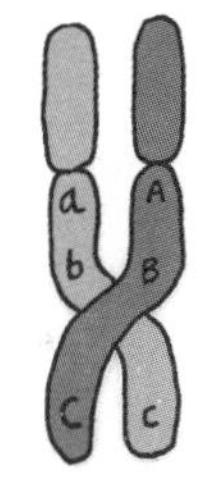

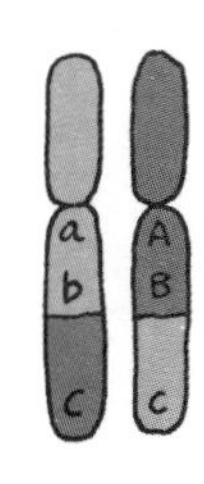

1931
Harriet Creighton (1909–2004) and Barbara McClintock (1902–92) show physical basis of crossing-over

1932
Muller describes usefulness of crossing-over for beating 'Muller's ratchet'

genetic relatedness

Recombination explains how much of your DNA you share with your family, and why you differ genetically from your siblings. Half of your genetic material comes equally from your mother and father, because you were conceived from gametes produced by each. But although you might think that 50 per cent of your DNA is common to your brothers and sisters, too, that is true only on average. The randomness of recombination means that it is theoretically possible, though statistically extremely unlikely, that you have inherited a wholly different set of alleles to your siblings.

intertwine, and then break at the points where they are twisted together. These segments then fuse with their neighbours, so that genes 'cross over' between the chromosomes. The result is a gamete with an entirely new, jumbled chromosome that is an amalgam of paternal and maternal genes.

This crossing-over means that while every gamete gets a copy of every gene, the combination of alleles it carries will be unique. A man's sperm will not have chromosomes that come wholly from his mother or from his father, like the ones in his somatic cells. It will have new ones that contain chunks of genetic material from each of his parents. Genes are thus forever being assembled in slightly different arrangements, and recombination can occasionally even merge them to create new genes. Some permutations and gene mergers may be beneficial, while others are damaging. It is another source of heritable variation on which natural selection can act.

Recombination also allows scientists to map where genes are placed on chromosomes, using the concept of linkage that was introduced in Chapter 3. As T.H. Morgan understood, genes that lie close together on chromosomes tend to be inherited together too, and crossing-over is the reason. Genes are not swapped individually between chromosomes, but as

> **'The "sexual" method of reading a book would be to buy two copies, rip the pages out, and make a new copy by combining half the pages from one and half from the other, tossing a coin at each page to decide which original to take the page from and which to throw away.'**
>
> **Mark Ridley**

parts of larger blocks. If two genes are contained in the same block or 'haplotype', they will be linked – individuals who get one will tend to get the other as well.

A reason for sex In species that reproduce sexually, meiosis and recombination give every individual a genotype of its own, and this extra variation can be adaptive. In asexual reproduction, mutations are invariably transmitted to offspring, even if they are deleterious. This leads to an effect known as Muller's ratchet (see box), by which genomes tend to deteriorate in quality over time. Sex, through crossing-over, allows offspring to differ from their parents. Half of them will miss out on rogue genes that would have been passed on asexually, giving the species an advantage.

The genetic variety that sex engenders, too, means that germs and parasites find it harder to spread through entire populations at once. Such diversity makes it more likely that some individuals will have a degree of genetic resistance, so that some will always survive new epidemics, and breed future generations with some immunity. Sexual variety gives the species that enjoy it a head start in life.

Muller's ratchet

When an organism reproduces asexually, its entire genome will be copied into its offspring. Hermann Muller realized that this has a big drawback: if a copying error causes a deleterious mutation, it will always be passed on to all that individual's descendents. The same thing will happen each time new mutations arise, so that the organism's overall genetic quality deteriorates over time. Muller likened the process to a ratchet, with teeth that allow movement in one direction only.

Sex and recombination circumvent Muller's ratchet, as they mean that not every mutation in a parent is passed to its children. Many asexual organisms, such as bacteria, have evolved other means of swapping genes to avoid negative effects.

the condensed idea
Sex makes individuals genetically unique

07 Genes, proteins and DNA

Francis Crick: 'Once the central and unique role of proteins is admitted, there seems little point in genes doing anything else.'

It is rather distressing to pass urine that turns pitch black on exposure to the air, yet the condition that causes this, alkaptonuria, was little studied for centuries as it is largely harmless. In the 1890s, it caught the attention of Archibald Garrod, an English physician. When Mendel's ideas were rediscovered soon afterwards, Garrod noticed that this disorder followed a Mendelian pattern of inheritance. He identified not only one of the first diseases confirmed to have a genetic origin, but also a rule of the field as a whole: that genes work by producing proteins.

Though alkaptonuria is rare, affecting about one in 200,000 people, Garrod noted it was much more common following marriages between first cousins, and that in susceptible families the ratio of unaffected to affected children was almost exactly three to one. This, he realized, was precisely what would be expected if alkaptonuria were caused by a recessive gene, and not by infection as was commonly assumed.

Garrod's knowledge of biochemistry, too, led him to propose a function for that gene. What blackens the urine in alkaptonuria is a substance called homogentisic acid, which the body usually breaks down. Garrod suspected, correctly, that patients with the condition lacked an enzyme (a protein that catalyses chemical reactions) that was critical to its elimination. The result was that the chemical was excreted in urine, turning it black.

1869
Friedrich Miescher (1844–95) discovers DNA

1896
Archibald Garrod (1857–1936) begins studying origins of alkaptonuria

One gene, one protein Garrod deduced from these observations that the function of genes was to make proteins. Many other medical problems might be caused by similar 'inborn errors of metabolism', as he called them in the title of a 1909 book. It was a hugely significant insight, providing a means by which genes and genetic mutations influence biology. Perhaps because of the relative obscurity of the diseases he studied, however, Garrod's theories, like Mendel's, went unremarked for decades.

They also lacked direct evidence. That was to be supplied in the 1940s by George Beadle – another pupil of T.H. Morgan's – and American geneticist Edward Tatum. Beadle's work with fruit flies had suggested that eye colour might be fixed by chemical reactions under the control of genes, but the organism was too complex to prove the theory experimentally. Beadle and Tatum turned instead to a simple bread mould called *Neurospora crassa*, which was irradiated to generate mutations.

When the mutants were crossed with normal mould, some of their offspring multiplied freely, but others would divide only when a specific amino acid, arginine, was added to the growth medium. These moulds had inherited a mutation in the gene for an enzyme critical to arginine production. Unless the essential amino acid was provided from elsewhere, the yeast could not grow.

This suggested a simple rule: every gene contains the instructions for making a particular enzyme, and that enzyme then goes to work in cells. Even though the rule has since been modified – some genes are capable of making more than one enzyme, or smaller components of proteins – it is essentially correct. Genes do not guide cellular chemistry directly, but by proxy through the proteins they make, or fail to make because of mutations.

It is an insight that has had profound implications for medicine: while altering the defective genes that cause disease is difficult, some genetic conditions can be treated by the more straightforward means of replacing the missing protein. Haemophiliacs, for example, can be given the blood-clotting enzyme that their bodies are genetically incapable of producing themselves.

1941

George Beadle (1903–89) and Edward Tatum (1909–75) confirm that genes make proteins and propose one-gene, one-enzyme hypothesis

1944

Oswald Avery (1877–1955), Maclyn McCarty (1911–2005) and Colin MacLeod (1909–72) show DNA carries genetic information

1952

Alfred Hershey (1908–77) and Martha Chase (1927–2003) use radioactive tagging to confirm DNA's genetic role

Life on Mars?

If primitive life is ever discovered on Mars – or anywhere else for that matter – the first question scientists will ask will be: 'is it based on DNA?' The genetic instructions of every terrestrial organism are written in DNA (the exception is some RNA viruses, and they cannot reproduce without a DNA-based host). That offers overwhelming evidence that they are all ultimately descended from a common ancestor.

If extraterrestrial life uses DNA as well, the same implication holds true. Perhaps Mars was seeded with life by microorganisms carried from Earth on a meteorite. Or perhaps the reverse is true – and we are all really Martians.

Enter DNA The discovery that genes carry the code for making proteins challenged conventional wisdom about their construction, as it had been widely thought that genes *were* proteins. If proteins were actually the products of genes, the chemical basis of heredity had to lie elsewhere. It was to be found in a mysterious substance that had first been purified from pus-soaked bandages by Swiss scientist Friedrich Miescher as long ago as 1869: deoxyribonucleic acid, or DNA.

DNA was known to exist in almost every kind of cell, yet though Miescher had suspected it might play a part in inheritance, this function remained purely speculative until Oswald Avery, Maclyn McCarty and Colin MacLeod began an important series of experiments in 1928. Avery's team was intrigued by a bacterium that causes pneumonia, which exists in two forms that are either lethal or harmless. When the scientists injected mice with both live harmless bacteria, and lethal ones that had been inactivated, they were surprised to see the rodents fall ill and die. The harmless germs had somehow acquired the virulence of the inert ones.

To find what they called the 'transforming factor', the scientists experimented with more than 20 gallons of bacteria for more than a decade. They treated these colonies with enzyme after enzyme that knock out particular chemicals, to test various candidates that might be conveying lethal instructions from germ to germ. Only when an enzyme that breaks down DNA was tried did the transformation stop: DNA was the messenger. Further evidence for its role came in 1952 from Alfred

Hershey and Martha Chase, who tagged DNA with radiation to show that it is the genetic material of a phage – a kind of virus that attacks bacteria.

DNA is not just the stuff of life for bacteria and phages: it writes the genetic recipes for every living thing on Earth. The only exception is certain viruses that use its chemical cousin ribonucleic acid (RNA) instead – and as these cannot reproduce on their own, there is some debate about whether they can really be considered to be alive.

The DNA code is written in only four 'letters' known as nucleotides or bases (see box). Yet this simple alphabet is sufficient to make organisms as different as humans and herring, frogs and ferns. It builds both the genes that produce proteins and the genetic switches that turn them on and off, and it is self-replicating, so the whole code can be copied every time a cell divides. It is life's software, containing the information needed to build and run a body.

The alphabet of DNA

Each DNA molecule is made up of phosphates and sugars, which provide its structural architecture, and nitrogen-rich chemicals known as nucleotides or bases, which encode genetic information. The bases come in four varieties – adenine (A), cytosine (C), guanine (G) and thymine (T) – and together, these provide the letters in which the genetic code is written.

The bases can be further subdivided into two classes: adenine and guanine are larger structures called purines, and cytosine and thymine are smaller pyramidines. Each purine has a complementary pyramidine, to which it will bind – A binds to T, and C to G. Mutations also tend to substitute a purine for a purine, or a pyramidine for a pyramidine – A will usually mutate to G, and C to T.

the condensed idea
Genes make proteins, and are made of DNA

08 The double helix

James Watson: 'At that time . . . all I cared about was the DNA structure . . . It was aided by the fact that effectively there were no girls in Cambridge.'

When Francis Crick sat down to lunch with James Watson at the *Eagle* pub in Cambridge on 28 February 1953, and announced that the pair had 'found the secret of life', other drinkers could have been forgiven a little scepticism. Crick was a 36-year-old physicist who had yet to finish his PhD. His American collaborator was just 24. They had also been expressly forbidden from studying the problem they now claimed to have solved: the structure of the DNA molecule that had been recognized for almost a decade as the conveyor of heredity. Even Watson, not a man known for circumspection, was somewhat disconcerted by his friend's boldness, as he was still concerned that their answer might be wrong.

'It seems likely that most if not all the genetic information in any organism is carried by nucleic acid – usually by DNA, although certain small viruses use RNA as their genetic material.'

Francis Crick

He need not have worried. Their discovery that DNA is wound into a double helix ranks as one of the most important scientific achievements of the 20th century, on a par with Einstein's relativity and the splitting of the atom. While early genetics had shown clearly that genes drive inheritance, it had had little to say about the chemical processes involved. Crick and Watson changed that, demonstrating how genes actually work. They opened a new era of molecular biology, in which genetic activity could be tracked, charted and ultimately even changed.

timeline

1950
Erwin Chargaff (1905–2002) discovers that ratios of adenine to thymine and cytosine to guanine are always the same, suggesting that bases are paired

1951
Linus Pauling (1901–94) proposes a triple-helix structure for DNA

The idea of the double helix also pointed firmly to the route by which the code of life is copied as cells divide, with each strand providing a template from which genetic instructions can be duplicated. As Crick and Watson put it in the short paper they published in *Nature* that April: 'It has not escaped our notice that the specific pairing we have postulated immediately suggests a possible copying mechanism for the genetic material.'

The search for the structure The significance of DNA to heredity was widely suspected by the early 1950s, and several teams were seeking to resolve the molecule's structure. In the US, Linus Pauling had already shown that many proteins were coiled into a spring-like helix, and proposed, incorrectly, a triple helix for DNA. At King's College London, meanwhile, Rosalind Franklin and Maurice Wilkins were studying DNA using X-ray diffraction, which analyses how molecules scatter radiation for clues to their form.

In Cambridge, Crick and Watson were meant to be using the same tool for different purposes – Crick's target was protein structure, and Watson's a tobacco virus – but they found DNA more interesting. For a period, however, they were told not to study it by Laurence Bragg, their laboratory head, who felt it would be distracting and ill-mannered for them to stray into the same territory as King's.

The dark lady of DNA

Rosalind Franklin's role in the discovery of the double helix remains a source of great controversy. The importance of her X-ray images is beyond dispute, and observers such as Brenda Maddox, her biographer, have argued that she was a victim of sexism who has never been given the credit she deserves.

Crick, Wilkins and, particularly, Watson, certainly failed to acknowledge her contribution properly at the time, but there is some merit in their counterargument that while Franklin's work was pivotal, she never properly grasped its significance. She was also excluded from the Nobel Prize for Medicine that the trio shared in 1962 for a perfectly innocent reason: she died of ovarian cancer in 1958, and Nobel Prizes are never awarded posthumously.

1952

Rosalind Franklin (1920–58) takes X-ray crystallography image of DNA that suggests a double helix

1953

Francis Crick (1916–2004) and James Watson (b.1928) identify the double helix

Linus Pauling

In the race to identify the structure of DNA, the smart money was not on Watson and Crick but on Linus Pauling – the brilliant American chemist who had already made key discoveries about protein structure and chemical bonding. Pauling was the first to suggest a helical structure for the DNA molecule, even if it was wrong in several details, and he might well have beaten the Cambridge team to the punch had it not been for his political activism.

In 1952, he was accused of having communist sympathies, and his passport was suspended. He was thus forced to abandon a trip to the UK, and never got to see the images taken by Franklin that helped Watson and Crick to solve the problem.

They continued to work on the problem, at first surreptitiously and eventually with Bragg's approval, and solved it by combining others' work with their own through luck, brilliance and deviousness. Their first stroke of fortune came from a 1952 visit to the UK by Erwin Chargaff, whose experiments in the US had shown that the four bases of DNA always occur in the same ratios – cells have equal amounts of the base pairs adenine (A) and thymine (T), and cytosine (C) and guanine (G). His lectures led Crick and Watson to understand that DNA bases come in pairs, with the letter A always linked to T and C linked to G. A critical piece of the double helix was in place.

A second vital clue came from Franklin's research. In 1952, she had taken an X-ray image of the DNA molecule, known as Photo 51, which Wilkins had shown to Watson without her knowledge. Crick, too, had learned of her results from Max Perutz, his thesis supervisor, who had reviewed the King's work for the Medical Research Council. The pair realized that the picture's significance had passed their rivals by, and that in combination with Chargaff's ratios it suggested a potential structure for DNA.

They were then able to turn this insight into results because, unlike Franklin, they did not confine their investigations to the laboratory. While the X-ray image was crucial, Crick and Watson understood its meaning by lower-tech means, playing with tin and cardboard models of DNA's components to test possible structures by trial and error. Photo 51 worked like the key to a jigsaw puzzle, indicating a framework into which all the pieces might fit. And the framework – the double helix – worked perfectly.

How the helix works The DNA molecule is composed of two linked chains of bases. Each base is joined to its natural partner – A to T and C to G – by a hydrogen bond, and held at the other end by a sugar and phosphate backbone. This pairing system means that two DNA strands coil around each other in a double helix, like a twisted rope ladder. Each strand is the mirror image of the other – where one has an A, its partner will always have a T, and vice versa. If the first strand reads ACGTTACCGTC, the other will read TGCAATGGCAG.

Double helix

This structure betrays its function. The sequence of the DNA bases encodes genetic information twice over, making it wonderfully easy to copy. When a cell divides, an enzyme breaks the hydrogen bonds that connect the base pairs, unzipping the double helix down the middle into its two constituent strands. These can then serve as templates for replication. A second enzyme called DNA polymerase tacks new bases onto the letters of each strand, matching As to Ts and Cs to Gs. The result is two new double strands of DNA, to provide the genetic software for two daughter cells.

Replication

1 Double helix unzips during cell division
2 Each independent strand of DNA acts as a template for a complementary strand to be created, adding As to Ts, Cs to Gs, etc.
3. Two new double-stranded DNA molecules are created, one of which migrates into each new cell

Like so many great ideas in genetics, the double helix is elegantly simple. Yet it immediately explained how the code of life is copied, and it cleared the way for further discoveries about how that code influences biology. It was the harbinger of a new genetic age in which it was to become possible to use DNA to diagnose disease, to develop drugs, to catch criminals and even to modify life. If the structure proved simple, the same cannot be said for its consequences.

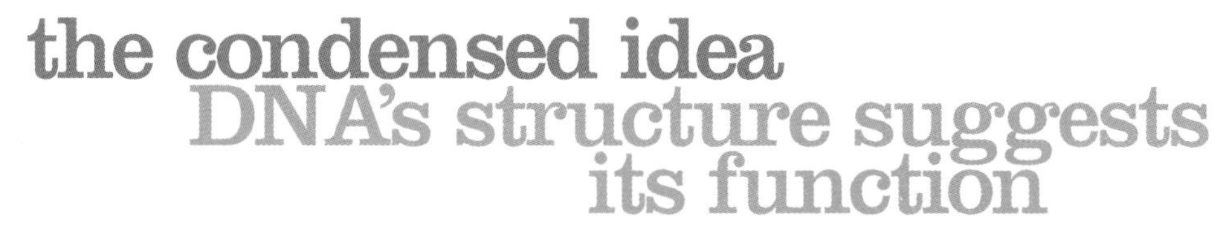

09 Cracking the code of life

Francis Crick: 'It now seems very likely that many of the 64 triplets, possibly most of them, may code one amino acid or another, and that in general several distinct triplets may code one amino acid.'

The double helix explained how genes are copied, and thus how genetic information is reliably passed on from cell to cell and from generation to generation. It also suggested that mutations in DNA's letters would be inherited, allowing for Darwin's descent with modification. What the structure did not elucidate, however, was how genes perform their other vital task besides copying themselves: the synthesis of the proteins that drive biology.

The code of life was clearly written in an alphabet of four letters – the A, C, G and T of DNA – which composed instructions for producing the 20 amino acids that make up proteins. But until it could be cracked, the code was meaningless. Biology had no Rosetta Stone, no key for deciphering the messages encrypted in DNA.

That key was to be provided by some perceptive theorizing by Francis Crick, followed by experiments led by American biochemist Marshall Nirenberg and French biologist Jacques Monod that found evidence to match. Little more than a decade after its discovery, written into the double helix, the genetic code had been broken and molecular biology had an organizing principle.

The adaptor molecule: messenger RNA Crick and Watson had identified the double helix by assembling existing evidence and interpreting it correctly. Crick's next stroke of genius, however, was much

1958

Crick proposes triplet coding system for DNA, adaptor molecule role of RNA, and the 'central dogma'

more speculative, emerging well in advance of any experimental data. It was the notion that DNA might be translated into amino acids by means of an 'adaptor molecule' – an interpreter that carries orders from genes to cellular protein factories.

By 1960, Crick's intuition had been proved right. At the Pasteur Institute in Paris, Monod's team used bacteria and bacteriophage viruses that prey on them to show that DNA indeed produces an adaptor molecule, which is made out of a close chemical relative called ribonucleic acid (RNA).

RNA is similar to DNA, but has a few structural differences. The most important is that instead of the base thymine, it uses a similar nucleotide called uracil (U). It is also more unstable, and thus shorter lived in the cell, and it forms many different types of molecules with specialized functions. Crick's adaptor molecule is a form known as messenger RNA (mRNA), a single-stranded molecule into which genes are 'transcribed'. This mRNA is used to make protein, in a process known as translation.

The central dogma

Another of Francis Crick's important contributions was what he called the 'central dogma' of biology – that genetic information generally travels through a one-way system. DNA can copy itself into DNA or transcribe itself into mRNA, and mRNA can make protein, but it isn't possible to reverse the process.

There are three exceptions to this rule. Some viruses can replicate themselves by copying RNA directly into RNA, or perform 'reverse transcription' from RNA into DNA. It is also possible to translate DNA directly into protein, but only in the laboratory. The information contained in proteins, however, can never be converted into RNA, DNA or even other proteins. The redundancy of the genetic code makes this impossible.

1960
Jacques Monod (1910–76) proves messenger RNA is the adaptor molecule

1961
Marshall Nirenberg (b.1927) discovers first triplet code for an amino acid

1966
Full range of 64 triplets identified

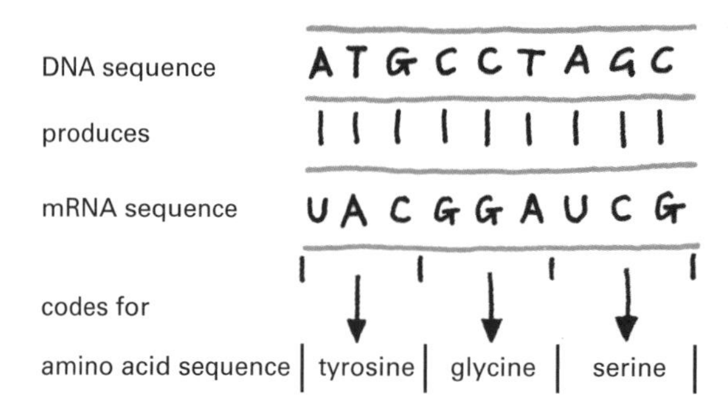

As in replication, the double helix is unzipped, and then one strand is read to produce a mirror-image strand of mRNA. In this transcription, Cs in the genes will become Gs in the mRNA, Ts will become As, Gs will become Cs, and As will become Us – uracil replacing thymine in the RNA molecule. These genetic signals then migrate from the cell nucleus to protein-making structures called ribosomes, which add amino acids into chains one at a time in the order specified by the code. Another kind of RNA, transfer RNA, collects amino acids and threads them onto the growing protein string. The instability of mRNA means that as in *Mission: Impossible*, the messages self-destruct when they have done their job. There is no danger that they will stick around to create rogue proteins when they are no longer needed.

Triplets How, though, do the ribosomes know which amino acids have been ordered? And how do they know where protein chains should start and finish? The answer lies in the sequence of the bases in genes, by which passages of DNA and mRNA specify particular amino acids. The code – first proposed by Crick – is extremely simple, based on combinations of just three DNA letters, or 'triplets'.

Exons and introns

Not all the DNA in a gene is actually used to make protein. The parts that matter are known as exons. These are interspersed with stretches of non-coding DNA called introns, which have no bearing on the protein recipe that the gene carries. While all the DNA is copied into mRNA, the introns are then edited out by special enzymes, and the exons are spliced together to place an order for a protein.

A good analogy is a film screened on television. The scenes you want to watch are the exons, but these are broken up by commercial breaks – the introns – which aren't part of the story. And if you have recorded the whole thing, you can fast-forward through the adverts to watch it all at once – much as a ribosome reads a spliced string of exons.

The meaning of these triplets began to emerge with the work of Nirenberg, who in 1961 mixed ribosomes from *E. coli* bacteria with amino acids and single RNA bases. When he added pure uracil, the result was long protein-like chains of the amino acid phenylalanine. The first triplet had been deciphered – mRNA bearing the message 'UUU' means 'put a phenylalanine molecule on the protein string'. Within five years, the meaning of all 64 combinations of the four bases had been established. The code had been cracked.

As there are 64 possible triplets or 'codons', yet only 20 amino acids, some amino acids are specified by more than one. Phenylalanine, for example, is made not only by the codon UUU, but also by the codon UUC. There are six ways of making each of the amino acids leucine, serine and arginine. Only two of the 20 are specified by unique codons: tryptophan (UGG) and methionine (AUG). AUG is also the 'start codon', telling ribosomes to begin adding amino acids, meaning that most proteins start with methionine. And there are three 'stop codons' – UAA, UAG, or UGA tell the ribosome: 'this protein is now complete'.

This system is not as simple as it might be. Crick, indeed, first proposed a more elegant code of 20 possible triplets – one for each amino acid. What nature's version lacks in style, though, it makes up for in substance, for its redundancy has considerable advantages. The fact that the most important amino acids can be produced by multiple codons creates resistance to mutation. Glycine, for example, can be written GGA, GGC, GGG or GGU. If the final base mutates, the product remains the same.

'It seems virtually certain that a single chain of RNA can act as messenger RNA.'

Francis Crick

This leaves less scope for catastrophic copying errors that could compromise a whole organism. About a quarter of all possible mutations, indeed, are 'synonymous' in this way, and the workings of natural selection mean that a still larger proportion of those that survive – about 75 per cent – have no effect on a protein's function. The genetic code is a 'Goldilocks' language – the amount of variation it allows is neither too much or too little, but just right for evolution.

the condensed idea
The code is written in triplets

10 Genetic engineering

Jeremy Rifkin: 'What the public needs to understand is that these new technologies, especially recombinant DNA technology, allow scientists to bypass biological boundaries altogether.'

A good code isn't just there to be deciphered and read. It can be used for creative purposes, too. If the code of life could be understood, it could potentially be altered and manipulated.

Hermann Muller had realized, when exposing his fruit flies to radiation in the 1920s, that deliberately inducing mutations might allow humanity to direct evolution in desirable ways. The double helix and the cracking of the genetic code meant this might be achieved with precision: instead of waiting for a random X-ray mutation with a useful function, perhaps chromosomes and genes could be edited with specific functions in mind. Genetic engineering was now on the agenda.

It is one thing to imagine genetic engineering, however, and quite another to perform it. An engineer of any kind is nothing without tools, and DNA triplets can hardly be cut and pasted with scissors and glue. What turned genetic engineering from science fiction into reality was the discovery in the 1970s of molecular 'scissors and glue', a series of enzymes that could be used for writing and copying genes, cutting them out, and splicing them into a genome. Scientists would now be able to play God, creating new combinations of DNA that had never before existed in nature.

Molecular scissors The first tools to be discovered were a class of proteins called restriction enzymes, which are sometimes nicknamed

timeline

1927
Hermann Muller (1890–1967) suggests that the genetic code could be deliberately manipulated by inducing mutations

1960s
Werner Arber (b.1929) discovers restriction enzymes

molecular scissors. Bacteria use these chemicals to 'restrict' bacteriophage viruses that infect them, by recognizing particular sequences of their invaders' DNA and chopping them up at those points.

The process, first described by Swiss microbiologist Werner Arber in the 1950s, had obvious potential in genetics. If such enzymes target specific stretches of DNA, they could be used to cleave it into fragments at specific spots. In 1972, American microbiologist Hamilton Smith identified a restriction enzyme produced by the bacterium *Haemophilus influenzae* that did just that, attacking a phage at the same stretch of six base pairs every time.

More than 3,000 restriction enzymes are now known, each specialized to a particular DNA sequence. They are fundamental to genetic engineering, allowing scientists to splice genes and parts of genes. If a gene is known to start with one DNA sequence and to finish with another, the two restriction enzymes specific to those sequences can be used to cut it out.

Recombinant DNA Once restriction enzymes have been used to cleave DNA, enzymes called ligases can be used to stick it back together. If restriction enzymes are molecular scissors, ligases are molecular solder or glue. Various cut fragments can thus be joined to one another, or stitched

Reverse transcriptase

Another enzyme that has proved important to genetic research is reverse transcriptase, which was discovered by David Baltimore and Howard Temin in 1970. It is used by retroviruses such as HIV to transcribe their RNA code into DNA, and then insert it into cells so they can replicate. Many drugs for treating HIV and other viruses work by inhibiting reverse transcriptase.

The enzyme also makes it possible to turn messenger RNA into DNA in the laboratory. This can be a valuable tool for gene-hunting, allowing scientists to find transcribed mRNA messages, and use these to infer the DNA sequences from which they are derived.

1970
Hamilton Smith (b.1931) discovers first site-specific restriction enzyme

1973
Herbert Boyer (b.1936) and Stanley Cohen (b.1935) found Genentech, the first biotechnology company to exploit genetic engineering

1975
Asilomar conference develops safety protocols for recombinant DNA research

into the genome of another organism. The result is known as recombinant DNA – a sequence that has been pieced together by recombining segments in the laboratory.

Recombinant DNA was first created in the 1970s by American biochemist Paul Berg, who stuck together parts of a monkey virus called SV40 and a bacteriophage. His original plan was to insert this genetically modified virus into *E. coli* bacteria to allow it to replicate, but he paused. SV40 is harmless to humans, but what if genetic engineering were to change that? SV40 was known to promote tumour growth in mice, and *E. coli* inhabit the human gut. If bacteria carrying the recombinant virus were to escape, they might infect people and churn out carcinogenic SV40 proteins.

The potential biohazard led Berg to suspend this stage of his experiments, and to call for a moratorium on allowing recombinant DNA to replicate until the risks could be properly assessed. He resumed only in 1976, after the Asilomar conference had drawn up strict safety protocols for future research (see box). Similar issues have dogged genetic engineering ever since: though thousands of recombinant products have been used safely in the past three decades, many critics still argue for a precautionary approach.

The first genetically modified organisms Less squeamish in outlook – or more reckless, depending on your point of view – were Herbert Boyer, of the University of California at San Francisco, and Stanley Cohen, of Stanford University. When they teamed up, Boyer was studying restriction enzymes, while Cohen was investigating plasmids – circular packages of DNA found in bacteria, which they sometimes swap between themselves as a defence mechanism against antibiotics or phages. Boyer and Cohen used the new tools of genetic engineering to add a gene

The Asilomar conference

In February 1975, Paul Berg assembled 140 scientists, doctors and lawyers at the Asilomar State Beach conference centre in California, to discuss the ethical questions raised by genetic engineering. It established a number of biosafety principles, designed to prevent the accidental release of a recombinant organism that could infect humans or animals. The key recommendation was that when human or animal viruses were being studied, bacterial hosts that cannot survive outside the laboratory should be used. That way, there was little chance of unwittingly unleashing a 'superbug' on the world.

that confers antibiotic resistance to a plasmid, and then inserted it into *E. coli*. The bacteria became antibiotic-resistant. They were the first true genetically modified organisms (GMOs).

The first application of recombinant DNA was in the laboratory, to 'clone' interesting genes by cutting them out and splicing them into plasmids. When placed into bacteria, these plasmids would then replicate, making multiple copies of the genes for scientists to study. A variation on this procedure was used to clone the segments of the human genetic code that were mapped by the Human Genome Project (see Chapter 12).

'The concern of some that moving DNA between species would breach customary breeding barriers and have profound effects on natural evolutionary processes has substantially disappeared as the science revealed that such exchanges occur in nature.'

Paul Berg

Still more exciting – and more lucrative – was the medical potential. Boyer saw that if human genes were engineered into plasmids, it would be possible to induce bacteria to make human proteins that could be used in therapy. In 1976, he set up a company called Genentech to commercialize the technology, with backing from Robert Swanson, a venture capitalist.

The company's first success was a recombinant version of insulin (the hormone essential to sugar metabolism that is lacking in patients with type 1 diabetes) which had previously been obtained from pigs. Boyer created this by placing the human insulin gene into *E. coli* via a plasmid. The bacteria that took up the plasmid became insulin factories, producing vast quantities of the hormone suitable for medical use.

A similar approach is now used to create scores of drugs and other commercial products, many of which have significant advantages over the alternatives. Human growth hormone for treating dwarfism, for example, was once extracted from the pituitary glands of cadavers, and contamination infected many recipients with Creutzfeldt–Jakob disease – the human equivalent of mad cow disease. The recombinant version has no such risks. There is even a recombinant form of rennet for making vegetarian cheese – the real thing comes from cows' stomachs. Muller had been spot on: genes could be moulded to our purposes.

the condensed idea
Genes can be manipulated

11 Reading the genome

Fred Sanger: 'This theme [sequencing] has been at the centre of all my research since 1943, both because of its intrinsic fascination and my conviction that a knowledge of sequences could contribute much to our understanding of living matter.'

By the early 1970s, science understood the double-helix structure of DNA, the triplets by which it codes for proteins, and many of the sequences of amino acids from which these cellular workhorses are built. Hamilton Smith, Paul Berg and Herbert Boyer had also taken the first steps to genetic engineering, establishing how simple segments of DNA can be transferred from one organism to another.

Further progress in understanding genetics and exploiting it in medicine, however, was still hindered by a major technical barrier. It remained extremely difficult to figure out which bits of DNA worked as discrete genes, and the order of the DNA 'letters' in which they are written.

The first gene was isolated from a bacterium by American geneticist Jonathan Beckwith in 1969, and the first gene sequence, for the protein coat of a virus, was determined by Belgian molecular biologist Walter Fiers in 1972. These achievements, however, involved reading RNA copies of the genetic code, not the raw DNA itself. The technique was slow and inefficient, and as RNA is so short-lived, it was inappropriate for all but the smallest genes. There was no way of routinely reading the order of DNA bases, and thus little prospect of mapping complex genes, let alone the complete genetic sequences of large organisms.

timeline

1972
Walter Fiers (b.1931) determines first gene sequence

1975
Fred Sanger (b.1918) develops chain termination sequencing

A superior sequencing method was eventually developed in 1975 by Fred Sanger, a British biochemist who had already won one Nobel Prize for determining the amino acid sequence of insulin. The method changed the face of biology, transforming the prospects for understanding and manipulating genes' functions, and ultimately allowing scientists to map humanity's genetic code.

Genome sequencing Sanger's novel approach was to use a single strand of DNA as a template for four experiments in separate dishes. In each pot, he placed a soup of the four bases – A, C, G and T – and DNA polymerase, an enzyme that starts using these to make a new complementary strand. A 'magic ingredient' is then added to each experiment – a modified version of one of the bases, which stops the reaction as soon as it is used in the strand, and marks its end with a radioactive label.

The Nobel Prize

Only four people have been awarded a Nobel Prize twice, and two of them were honoured for their discoveries in genetics. Fred Sanger is a double laureate in chemistry, and Linus Pauling has won the chemistry and peace awards. The physiology or medicine prize, too, has come to be dominated by genetics, particularly since the science began to surge forward in the 1950s. The roll-call of laureates reads like a who's-who of the history of genetics: Morgan, Muller, Beadle, Tatum, Crick, Watson, Wilkins, Nirenberg, Monod, Smith, Baltimore and Cohen. Five of the last seven prizes have also rewarded discoveries with a genetic component.

As the reactions progress, they generate thousands of DNA fragments of varying lengths, some of which will end at every possible position. These are then forced through a gel, to separate them by size in order of length, and the base at the end of each piece can be read from its radioactive marker.

If the first fragments, with just one base, have thymine at the end, the first letter is a T. If the fragments with two bases have cytosine at the end, the code can be built up to TC. Three-base fragments with guanine at the end make the sequence TCG. Every piece is then read in the same way, until every place in the code has been filled in with a letter.

1977
Sanger sequences the first genome of an entire organism, a phage virus called Phi-X174

1981
Sanger's team sequences human mitochondrial genome

1991
Craig Venter (b.1946) develops fast new method for finding genes, using expressed sequence tags

This system, known as chain-termination sequencing, was much faster than the alternatives. It was efficient, reliable and safe – other new techniques that were developed at much the same time used more radioactivity and toxic chemicals. It rapidly became the gene-reading method of choice.

Chain termination sequencing

1 Single-stranded DNA sequence

ACGTGCCATTA

2 Single strands of DNA split into chunks of every possible length, and final base tagged with radiation

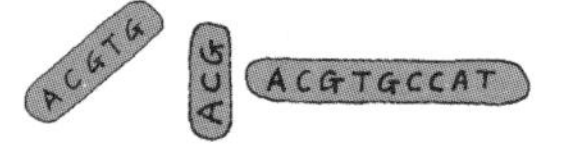

etc.

3 Radioactive tag at the end of each chunk is read, and chunks are lined up in order of length to generate the sequence

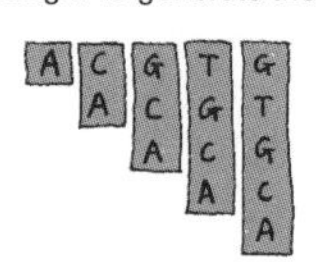

etc.

At first, it was done by hand. When Sanger used it to read the genome of a phage virus called Phi-X174 – the first DNA-based organism to be sequenced in full – he counted off the bases one-by-one from bands in a piece of gel. This process was obviously time-consuming and costly, but it was amenable to automation. In 1986, Leroy Hood, of the California Institute of Technology, invented the first DNA sequencing machine. Instead of using radioactivity to identify the bases, Hood labelled them with four fluorescent dyes, which glow when scanned with a laser. Each light signature is then identified by a computer, which steadily builds up the sequence. Technicians need examine no slides. Machines made by Applied Biosystems, the company that commercialized Hood's invention, were used to sequence the human genome.

Hunting genes These new sequencing techniques made reading the letters that make up genes much simpler. Finding genes in the first place, however, remained difficult. Scientists would first purify a protein such as adrenaline from cells, and then work out its amino acid sequence and all the possible combinations of DNA triplets in which its genetic instructions might be written. The process could take years.

From these candidate DNA sequences, it was possible to make a 'DNA probe' to seek them out in chromosomes by exploiting an aspect of the double helix discovered by Crick and Watson. Single strands of DNA will bind to other single strands made of complementary bases – a sequence that reads ACGT will stick to one that reads TGCA. A DNA probe bearing part of the sequence of the candidate gene could be labelled radioactively, and then mixed with genetic material from chromosomes. If it stuck to anything, that was probably the real gene, which could then be isolated, read and mapped to its position on its chromosome.

By the end of the 1980s, almost 2,000 genes had been found in this way and then sequenced. One of these codes was for erythropoietin – a protein that boosts red blood cell production. When a company called Amgen developed a recombinant version, it became a blockbuster drug that transformed treatment of anaemia. But despite heavy investment from the pharmaceutical industry, which thought more money-spinners were just around the corner, the pace of discovery remained slow.

That pace was suddenly to pick up in the early 1990s, thanks to a new gene-hunting technique devised by Craig Venter, a Californian surfer who had come late to biology after serving as a medical orderly in Vietnam. He realized that by sequencing small sections of DNA that are known to be copied into messenger RNA – the signalling chemical that builds proteins – it was possible to create 'expressed sequence tags' with which to fish for whole genes in chromosomal DNA. Armed with this method, his lab was soon discovering up to 60 new genes each day. The genome was starting to give up its secrets.

The first human genome project: mitochondrial DNA

The human genome is three billion bases long, and reading it was well beyond the sequencing tools available to Sanger in the late 1970s. That, however, did not stop him from embarking on a more limited kind of human genome project. While most human DNA is carried in the chromosomes of the cell's nucleus, a tiny amount exists in energy-producing structures called mitochondria. Sanger's team set about sequencing this more manageable chunk of our species' genetic code – and in 1981 published details of its 16,569 bases and 37 genes.

The mitochondria might be small, but they are not unimportant. Defects in mitochondrial genes can cause diseases, and scientists are currently investigating how to transplant them between eggs to stop these disorders from being inherited. As mitochondria are passed down the female line relatively intact, the DNA they contain is also useful for studying evolution and genealogy.

the condensed idea
Genes can be isolated and read

12 The human genome

John Sulston: 'The only reasonable way of dealing with the human genome sequence is to say that it belongs to us all – it is the common heritage of humankind.'

As DNA sequencing began to reveal human genes in the 1980s, an even bigger prize started to edge into view. If science could learn so much about biology and disease by mapping a few short sections of DNA, how much more might be revealed by reading the entire genetic code of our species?

While sequencing had to be conducted by hand, a project to decipher the whole human genome had looked like fantasy. But with the advent of automated techniques, influential figures started to argue that it might be both possible and worthwhile. In 1986, Renato Dulbecco, a Nobel laureate, called on the US government to support such an effort to underpin cancer research. In Britain, Sydney Brenner – a future Nobel prizewinner – was urging the European Union to do the same thing.

The US Department of Energy, which had been tasked with investigating the effects of radiation on DNA, soon took up the baton. 'Knowledge of the human genome is as necessary to the continuing progress of medicine and other health sciences as knowledge of human anatomy has been for the present state of medicine,' it declared in a 1986 report. But other scientists and institutions, including the US National Institutes of Health, were more sceptical. Some considered the task too ambitious and expensive. Others thought it would divert both intellectual and financial capital away from more achievable genetic research.

timeline

1986
Renato Dulbecco (b.1914) suggests sequencing the human genome to improve understanding of cancer

1990
Launch of international Human Genome Project

The Human Genome Project By the end of the decade, the case had been made. The international Human Genome Project, funded by governments and charities, was launched in 1990, under the leadership of James Watson. Its goal was to read every one of the 3 billion base pairs in which humanity's genetic instructions are written, which its architects envisaged would take 15 years and $3bn – a dollar per DNA letter.

The project was so grand in scope that the last thing that it expected was competition. Yet in 1998, when the public consortium had finished just 3 per cent of the code, a private-sector challenge emerged. Craig Venter, the geneticist who had identified more genes than any other, struck a $300m deal with the main manufacturer of DNA sequencing machines to produce his own version of the genome.

Armed with a new technique he had developed, called whole-genome shotgun sequencing, Venter's company, Celera, promised to finish in just two to three years, well before the public project's planned completion date. Watson had won the first great contest of the genetic era, to find the structure of DNA. Now another race was on, which was to become one of the most bitter rivalries in modern science.

Whose genome?

Both the Human Genome Project and Celera used genetic material from several donors: DNA was extracted from blood taken from women, and from sperm provided by men. Celera's genome used five individuals – two white men and three women of African-American, Chinese and Hispanic backgrounds. The male donors, it later emerged, were Craig Venter and Hamilton Smith. The public project used DNA from two males and two females. Though they remain anonymous, one is known to have been a man from Buffalo, New York, code-named RP11, whose sample was used most often because it was of the best quality.

1998
Venter's company, Celera, launches private sequencing effort

2000
Celera and Human Genome Project announce completion

2003
'Finished' human genome sequence published

Different techniques, different philosophies The human genome is much too big to be read in one go. It would have to be split into sections that sequencing machines could manage, and the rival teams took different approaches to this problem. The public project first divided it into large chunks of 150,000 base pairs, cloned thousands of copies in bacteria, and then mapped these clones' positions on their chromosomes. Each clone was then broken down still further into random fragments, sequenced and reassembled by matching the fragments' overlapping ends. The sequenced clones would then be mapped back to their chromosomal positions to provide the complete code.

The technique was thorough, but extremely slow. Celera, which took its name from the Latin for 'speed', thought it could do better by skipping the mapping stage, and assembling the whole genome at once from small fragments. This 'shotgun' method had already been used to sequence bacteria and viruses, but many experts doubted it would work on the human genome, which is larger by a factor of 500 or more. Venter, however, proved its utility by sequencing the genome of an old friend to geneticists – the fruit fly – and moved on to the human code.

Data release: a double-edged sword?

By publishing its new data on a daily basis, the Human Genome Project hoped to make it impossible for Celera to patent the entire genetic code. The strategy worked, but at a price. Celera was free to download the fruits of its rival's labours to refine its own sequence, and other biotechnology companies could and did do the same thing. As Craig Venter has pointed out, this probably led to more genetic patenting, not less, as businesses pored over the public results and filed claims on the most interesting-looking genes.

Had the two groups been divided by nothing more than professional approach, relations might have remained cordial. But they differed in world view, too. The Human Genome Project saw the genetic code as the universal property of humankind, and placed all results into a public database, GenBank, as soon as they were ready. Celera, however, was there to make money.

While Venter had forced his financial backers to accept that he would publish his data openly, he was also running a business. Celera hoped to sell access to a powerful genetic database, complete with software that companies could use to find new genes and develop new drugs. University researchers could see the data for free, but would have to pay royalties on any commercial products they might develop as a result.

This was anathema to scientists like John Sulston, who led the British contribution to the public project. They saw Venter as a kind of genetic pirate, who was trying to appropriate something that belonged to everybody for his investors and himself. Though Venter always insisted the genome itself was unpatentable, the fear was that Celera was seeking to privatize it. The Human Genome Project stepped up the pace of its efforts, hoping to block any bid for ownership by putting data into the public domain before its rival could lay a claim.

> **‘We are learning the language in which God created life.’**
>
> **Bill Clinton**

An agreed draw Venter finished first, but the public project was so close behind that he agreed to a draw. Critical to the uneasy truce were two interventions by Bill Clinton, the US president. In April 2000, Clinton announced that he thought the genome should be public property, and his statement sent biotech stock prices – including Celera's – into freefall. Mortified by this unintended consequence, he then determined to make amends by bringing the two parties together. He negotiated a joint announcement at the White House between the two camps, at which he formally acknowledged Venter's contribution.

Celera was true to its word and published, and its value-added database proved so useful that most public science institutions and pharmaceutical companies subscribed. The public project's gear-change, however, precluded any possibility of the genome being patented. In 2004, after Venter had fallen out with his backers and resigned from Celera, his reference genome was even added to GenBank, free from any access restrictions. The genome war was over, and the fierce rivalry had served humanity well. The stimulus of competition meant the genome had been sequenced far more quickly than anyone had thought possible a decade before.

the condensed idea
The genome belongs to everyone

13 Lessons of the genome

Tarjei Mikkelsen: 'Any distinctly human trait caused by DNA is caused by one or more of these 40 million genetic changes [between humans and chimpanzees].'

When the acrimonious race to read the human genome reached its final furlong, the competing parties could at least agree on one thing. The 'book of man' was going to contain an awful lot of genes.

The fruit fly, Craig Venter had shown, had around 13,500 genes. John Sulston's project to sequence *Caenorhabditis elegans*, a microscopic nematode worm, had revealed about 19,000. Human life was so complex, it was reasoned, that many more genes than that would be needed to write the instructions. The consensus figure was around 100,000, and one biotech company even claimed to have characterized 300,000 human genes.

The publication of the two draft genome sequences in 2001 was to deliver quite a surprise. Analysis suggested that it contained only between 30,000 and 40,000 genes, and the tally has fallen steadily ever since. At the time of writing, the latest count is about 21,500 – slightly more than the zebrafish, and slightly fewer than the mouse. There is little correlation between the biological complexity of an organism and its number of protein-coding genes.

One gene, many proteins Since the experiments of George Beadle and Edward Tatum in the 1940s established that genes make proteins, the notion that one gene codes for one protein had become a mantra of molecular biology. Yet there are hundreds of thousands of

timeline

1941
Beadle and Tatum show that genes make proteins

1961
Nirenberg discovers first triplet code for an amino acid

human proteins, but only tens of thousands of human genes. The mantra was wrong. Both genes and proteins, it turns out, are more versatile than had been assumed.

Single genes can in fact contain the recipes for many different proteins – in part because of their structure. As explained in Chapter 9, only the sections of genes known as exons actually carry instructions for protein synthesis. Information from non-coding introns is removed from messenger RNA, and the exons are stitched together, before proteins are made.

These exons can be spliced in many different ways, and this 'alternative splicing' means a single gene can specify multiple proteins. Some genes, too, make only chunks of proteins, which can then be joined together in different orders to produce a wider variety of enzymes. Proteins can also be modified by cells after they have been produced. The result of all these processes is a protein population or 'proteome' that is much more diverse than the human gene count would suggest.

The surprisingly low number of human genes also indicates that 'junk DNA' – the 97 to 98 per cent of the genome that does not code for protein – might be more important than had been imagined. Some non-coding regions produce different cellular messengers made out of specialized forms of RNA (see Chapter 48). These work as switches that turn gene activity on or off and up or down, or that direct splicing to change which protein a gene makes. Much junk DNA, indeed, is now thought to be anything but junk (see Chapter 45). Some of it is critical to regulating how genes are expressed, and is as physiologically significant as genes themselves.

Gene splicing

1 Gene's sequence

TACGCA AACCT GATCGA TCCG TACCAG

EXON 1 INTRON 1 EXON 2 INTRON 2 EXON 3

2 Whole gene transcibed into mRNA

AUGCGU UUGGA CUAGCU AGGC AUGGUC

3 Gene splicing removes introns, which do not carry protein-coding information

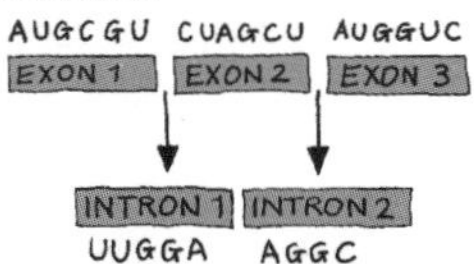

4 Exons, which carry protein-coding information, are translated into amino acids, which are then strung together into proteins

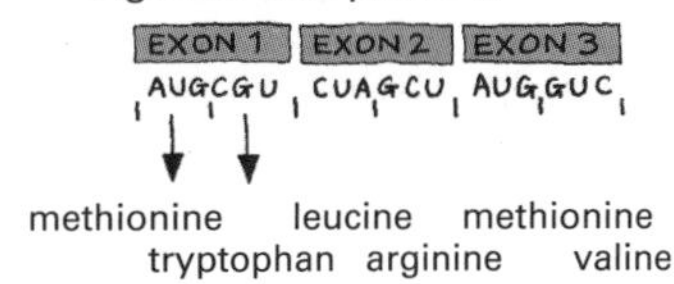

Variation between species When the human genome was compared with those of other species, it became clear that very few human genes are truly unique: most have a counterpart in other organisms.

1990s
Number of human genes estimated at 100,000 or more

2001
Human Genome Project shows total number of genes is no greater than 40,000

2008
Latest estimate of human gene count is 21,500

About 99 per cent are shared with chimpanzees, and about 97.5 per cent with mice. Natural selection does not reward change for change's sake, and thus genes that work well tend to be 'conserved' by evolution. A very similar code, making a very similar protein, will do the same job in related species. Both humans and pigs, for example, share a similar insulin gene: that is why pig insulin could be used to treat diabetics. Evolution does not often drop genes or create entirely new ones, so in retrospect, perhaps it is not so odd that most mammals have been found to have comparable gene counts.

What often happens instead is that a few genes are co-opted to perform new functions as evolution progresses. Many acquire slight mutations that are peculiar to a particular species, which allow them to do new things. A human gene called FOXP2, for example, has a counterpart in both mice and chimpanzees, but the human version differs from the chimp version in two places, and from the mouse version in three. These tiny changes may have played a role in the evolution of speech: people with defective FOXP2 genes suffer from language impairment.

How finished is the genome?

Most people tend to think the sequencing of the human genome was completed either in 2000, when the feat was announced at a White House press conference, or in 2001, when the rival groups first published their data. All that had then been produced, however, were working drafts that were riddled with holes: almost 20 per cent of the code had yet to be sequenced. Even in the supposedly 'finished' version released in 2003, about 1 per cent of the protein-coding regions was missing, along with higher proportions of non-coding junk DNA. Efforts to fill in the gaps are continuing, and the sequences of certain sections – the centromeres at the middle of chromosomes, and the telomeres at their ends – remain unmapped. They contain so much repetitive DNA that standard technology has struggled to read them.

Many of the differences between humans and other animals have emerged not because we have new genes unknown in our relatives, but because some of our shared genes have been altered, so they work in subtly different ways. Others are thought to reflect changes in the regulatory regions of junk DNA, and in the RNA messages that they send.

Variation between people People, of course, are genetically even more similar to one another than are humans and chimpanzees. By standard measures, 99.9 per cent of the genome sequence is universal, shared by every person on the planet. We also have the same genes as one another, except in rare cases in which one or more have been deleted entirely. The 0.1 per cent of DNA that is not shared, however, provides plenty of scope for variation: with 3 billion base pairs in the genome, that still leaves 3 million places in which individuals' DNA can differ.

This kind of variation involves random substitution of one DNA letter for another. The places where this occurs are known as single nucleotide polymorphisms or SNPs – pronounced 'snips'. Many SNPs have no effect at all: as discussed in Chapter 9, the genetic code has redundancy, so some mutations do not change the amino acid sequence of proteins. Others, however, make a material difference to the protein that a gene will produce, or alter the way junk DNA controls gene expression.

These SNPs are one of the primary ways in which genetics make individuals different. Some have trivial effects, altering characteristics such as hair or eye colour. Others are more insidious, either directly causing disease or altering metabolism in ways that make people more vulnerable to particular conditions. They are responsible for much of the variety of human life.

There is no human genome

When we talk about the human genome, we are in a sense discussing a fictional entity. The only people who share every letter of the genetic code are identical twins, and every one of us is otherwise unique. What the human genome sequence provides is an average, a reference point against which all of our individual genetic variations can be compared. It tells us where the important genes that we do share lie, making it simpler to investigate what they do. This means that when scientists find SNPs that seem to be linked to a disease, it is possible to trace them back to the genes in which they occur, revealing clues to their effects.

the condensed idea
Genetic variation isn't all about new genes

14 Genetic determinism

Francis Galton: 'It would be quite practicable to produce a highly-gifted race of men by judicious marriages during several consecutive generations.'

When his version of the human genome sequence was published in February 2001, Craig Venter attended a biotechnology conference in Lyon, France. In his keynote speech, he extolled it as a landmark in human understanding, not only because of what it explains about the significance of genetics, but for what it says about the field's limits as well.

As it contained so few genes, Venter said, the genome put paid to any notion that the behaviour, character and physiology of individual human beings are wholly determined by their genetic constitutions. 'We simply do not have enough genes for this idea of biological determinism to be right,' he said. 'The wonderful diversity of the human species is not hard-wired in our genetic code. Our environments are critical.'

Venter's logic was somewhat flawed – John Sulston, indeed, has accused him of making a 'bogus philosophical point'. It is quite true that the 30,000 to 40,000 genes that the genome was then thought to contain are utterly insufficient to hard-wire every human trait. But his implication – that three times as many genes would have been able to do this – is wrong. Both genetic *and* environmental factors are important in explaining the human condition, and the genome did not initially shed much light on the relative importance of each.

Venter's intention to discredit genetic determinism, however, is in itself worth noting. For since the inception of the science, genetics has

timeline

1860s
Francis Galton (1822–1911) develops ideas of promoting 'hereditary genius' through breeding

1883
Galton coins term 'eugenics' to describe his movement

commonly been misinterpreted to imply predestination, and the idea that people are prisoners of their genes has had dreadful social and scientific consequences.

Social Darwinism When Charles Darwin published *On The Origin of Species* in 1859, he avoided discussion of what evolution had to say about human behaviour, but it was not long before contemporaries tried to apply his theories to society. Figures such as Herbert Spencer, a philosopher who coined the phrase 'survival of the fittest', reasoned that human societies could learn from nature, and improve themselves by marginalizing and discarding their weakest members. These 'social Darwinists' argued that interventions to help the poor and the sickly might be noble in intention, but that they would ultimately weaken the human race by subverting natural selection.

Other thinkers appropriated Darwin to support their own notions of biological determinism. Cesare Lombroso (1836–1909) and Paul Broca (1824–80) contended that criminals, the mentally ill and the intellectually challenged were physiologically different from ordinary law-abiding citizens, and that their poor character was inherited and immutable. The now-discredited pseudosciences of phrenology and craniology, which hold that certain physical features and skull shapes reflect moral or mental degeneracy, were widely used to support such views.

Evolutionary theory was also employed to champion racism, with the argument that certain ethnic groups, particularly those with dark skin, represented more primitive forms of humanity that owed savagery to their less evolved status. Robert Knox, a Scottish anatomist, developed a particularly repellent anthropological theory, which contended that mankind was a genus, and that the different human races were species of greater or lesser sophistication, which could be scientifically classified in order of superiority. White Anglo-Saxons, of course, stood at the apex of his ethnic hierarchy.

'On the towpath we met and had to pass a long line of imbeciles. It was perfectly horrible. They should certainly be killed.'

Virginia Woolf

1912
UK Mental Deficiencies Bill withdrawn after campaign by Josiah Wedgwood

1927
US Supreme Court upholds compulsory sterilization laws in Buck v Bell case

1933
400,000 forced sterilizations in Nazi Germany

Galton the polymath

Francis Galton is usually remembered today for eugenics, but many of his other achievements were based on much sounder science, and proved more lasting. His experiments on rabbits showed that traits are not passed on by mixing parental characteristics, as Darwin had thought likely, and foreshadowed Mendelian genetics. He effectively founded modern statistics, introducing the principle of regression to the mean, by which abnormal results tend to be followed by a return to the average. He also helped to develop the forensic science of fingerprinting, and to advance meteorology: Galton produced the first weather map. His one bad idea has largely obscured his many good ones.

Eugenics Darwin rejected social theories that drew on his biology, influenced in part by his own family's history of medical infirmities: two of his ten children died in infancy, and he was particularly devastated by the death of his daughter Annie at the age of ten. His cousin Francis Galton, however, was to take up such thinking with fervour. A polymath of formidable intellect, Galton drew the conclusion from his research into human inheritance that the species could be improved by selective breeding, like any other species. He was the founder of eugenics.

‘And for the rest – those swarms of black and brown and yellow people who do not come into the new needs of efficiency? Well, the world is a world, not a charitable institution, and I take it they will have to go.’

H.G. Wells

This philosophy, which takes its name from the Greek words for ‘good breeding’, at first had the goal of producing a gifted elite caste by encouraging ‘eugenic marriages’ between people of good health and high intelligence. It soon took on a more sinister form, with advocates seeking to discourage or even to prevent reproduction among those deemed to come from lesser genetic stock. At worst, they promoted the forcible sterilization of ‘imbeciles’, the disabled, the insane and others considered to be genetically unfit.

In the late 19th and early 20th centuries both positive and negative eugenics were widely seen as progressive and scientific. Some of the movement's most enthusiastic supporters were socialists such as H.G. Wells and Beatrice and Sidney Webb, who saw it as a means of improving the genetic quality – and hence the social prospects – of the working classes.

Despite its British origins, eugenic measures were never incorporated into UK law (see box), but many other countries adopted them enthusiastically. In the US, many states passed

Britain's eugenics bill

In 1912, the UK's Liberal government introduced the Mental Deficiencies Bill, backed by no less a champion than Winston Churchill. It would have imposed penalties on people who married partners thought to be intellectually subnormal, and it was drafted so it could later be amended to approve compulsory sterilization. The campaign against it was led by Josiah Wedgwood, a Liberal MP who, like Galton, was related to Darwin. Wedgwood attacked both the shaky scientific principles on which the bill was based, and its assault on individual liberty, and won enough support to secure its withdrawal. It was the closest Britain came to eugenic legislation.

eugenic marriage laws that banned the 'feeble-minded' or even epileptics from marrying, and 64,000 people had been forcibly sterilized by the time the practice was finally outlawed in the 1970s. Nazi Germany went still further, progressing from 400,000 forced sterilizations in the name of 'racial hygiene', to euthanasia of the disabled, and ultimately to the Holocaust.

Multiple misunderstandings Leaving aside these appalling infringements of human liberty, the sort of biological determinism that drove the eugenics movement rested on a huge scientific misunderstanding. While genes have an important influence on many aspects of human health and behaviour, and many diseases and mental disorders are inherited, most of the traits and conditions that the eugenicists sought to affect are not governed by genetics alone. Venter was right in the broadest sense: genes by and large do not programme human behaviour and health, but exert an altogether subtler influence.

The crimes inspired by a warped misinterpretation of inheritance, however, have had a lasting impact on a whole field of study. Past abuses of genetics have left many people so suspicious of any suggestion that genes play a part in forming human character or behaviour, that even investigating such effects is often seen as politically incorrect. That is no more scientific than the mistaken theories of Galton or Knox.

the condensed idea
Genes influence, but rarely determine

15 Selfish genes

Richard Dawkins: 'We are survival machines . . . robot vehicles blindly programmed to preserve the selfish molecules known as genes. This is a truth which still fills me with astonishment.'

To many people, the 'bible' of genetic determinism was published in 1976 by Richard Dawkins, then a little-known Oxford University zoologist. Though *The Selfish Gene* contained little original research, drawing heavily on other scientists such as George Williams, William Hamilton and John Maynard Smith, it can justifiably be claimed as one of the most influential works in modern biology. It remains the outstanding account of the gene-centred view of evolution.

The Selfish Gene's argument is that many traditional accounts of evolution and genetics have got a fundamental principle the wrong way around. Organisms do not use genes to reproduce; rather, it is genes that exploit organisms to replicate and pass themselves on to another generation. The gene is the basic unit of natural selection. Evolution is best understood as acting on these self-copying packages of information, and not the creatures, plants or bacteria that carry them.

At one level, this is banal – since the modern evolutionary synthesis, it has been accepted that genetic variation is the raw material that allows evolution to take place. At another level, however, it is highly provocative. It suggests that the phenotypes that genes create have no inherent value: though these might improve the survival and reproduction of individuals, groups and species, they are not ultimately selected for this purpose. Such benefits are the incidental means by which genes secure their future. *The Selfish Gene* is the strongest possible interpretation of natural selection's amoral nature, suggesting there are few facets of behaviour or physiology in which genetic influences might not be found.

timeline

1859
Darwin publishes *On the Origin of Species*

1865
Mendel identifies laws of inheritance

Survival machines The short life spans of all living things mean that individuals are here today, gone tomorrow. Their genes, however, are functionally immortal – at least for as long as they can continue to duplicate themselves, and live again in another body. They do this by building 'survival machines' – Dawkins's elegant phrase for roses, amoebae, tigers and people, which ferry genes from one generation to the next.

The genes that thrive and succeed in making the most copies of themselves are the ones that build survival machines that are best adapted to their environments. Genes thus often have beneficial functions in the organisms that carry them: they instruct cells to produce adrenaline to aid flight from predators, insulin to metabolize sugar, or dopamine to run the brain. But these adaptations are nothing more than a by-product of Darwinian selection's action at a genetic level, where it rewards those genes that copy themselves most often.

This is what Dawkins meant by invoking his brilliant metaphor: to an outside observer, genes look as if they are behaving selfishly. Organisms breathe, feed and behave in certain ways because it suits the interests of their genes that they do so. It is a paradigm that explains many known

Memes

Perhaps the most truly original idea in *The Selfish Gene* is that cultural phenomena can be subject to a form of natural selection, in similar fashion to genes. Dawkins coined the term 'meme' to describe a unit of cultural information – such as a religion, song or anecdote – which is passed from person to person and which competes for popularity. Like genes, memes can mutate when people copy them incorrectly. Advantageous mutations, which make a meme more memorable, tend to thrive, while those that ruin its meaning die out. The concept is highly controversial: some philosophers think it elegant, but others find the analogy too neat, and lacking evidence in its support.

1953
Crick and Watson identify double-helix structure of DNA

1966
George Williams (b.1926) proposes gene-centred view of evolution

1976
Richard Dawkins (b.1941) publishes *The Selfish Gene*

phenomena in biology and medicine – including the matter of why we grow infirm as we get older and eventually die. From a gene's perspective, there is no point in building survival machines that last a lot longer than their purpose, which is to live long enough to breed and raise young, so that their genes can prosper all over again.

A misunderstood metaphor His choice of language, however, left Dawkins open to misinterpretation – often wilful – from critics who felt his theory was overly bleak, reductionist and deterministic. Genes, of course, are not conscious and have no intentions: they are not selfish in the way people can be. As Mary Midgley, a philosopher, said in a famous review: 'Genes cannot be selfish or unselfish, any more than atoms can be jealous, elephants abstract or biscuits teleological.' This line of argument, however, was a classic assault on a straw man. Dawkins had made it perfectly clear that genes are not actually selfish, but that they work in ways that make them appear so. The entire point of his hypothesis is that evolution is motiveless.

Another implication that is often wrongly drawn from the book is that if genes work selfishly, individuals must behave in this manner too. Yet as Dawkins had again explained, selfish genes do not necessarily generate selfish people. In fact, they offer a wealth of potential evolutionary explanations for altruism. Within families, in which many genes are shared, individuals have an obvious genetic motivation for helping others.

The naturalistic fallacy

A common misconception about Dawkins, and the evolutionary psychologists he has helped to inspire, is that selfish gene theory seeks to justify a dubious morality. This argument falls into an intellectual trap called the naturalistic fallacy. That something is natural does not make it right. If genes can promote violence or rape, to help them to propagate, that provides no justification for such crimes, as Dawkins makes perfectly clear. Indeed, we need to study such influences if we want to prevent them. 'Let us understand what our own selfish genes are up to, because we may then at least have a chance to upset their designs, something that no other species has ever aspired to do,' he says.

Mathematical biologists have also used game theory to show that selfish genes can thrive by making organisms co-operate to their greater shared benefit – a concept known as 'reciprocal altruism'.

Neither does selfish gene theory imply that organisms can be explained solely in terms of their genes, as critics such as Midgley seem to think. The gene-centred view of evolution is a reductionist theory, but it is not deterministic: it does not exclude environmental inputs. Individuals' phenotypes, Dawkins says, are always a product of both genes and their surroundings. That, indeed, is one of the main reasons why evolution does not act on phenotypes, which always differ between individuals and are thus destroyed by death, but does act on the longer-lasting and much less mutable genes.

'*The Selfish Gene* brought about a silent and almost immediate revolution in biology. The explanations made so much sense, the fundamental arguments were so clearly stated and derived completely from first principles, that it is hard to see after reading the book how the world could ever have been any different.'

Alan Grafen

Evolutionary psychology One effect of *The Selfish Gene* was to inspire a generation of biologists to think afresh about how genes affect human life, helping to shape not only our bodies but also our minds. The gene-centred view fed a burgeoning understanding that people are animals, that the brain is an evolved organ, and that its propensities have not escaped the influence of selfish genes working to promote their own survival.

This has been particularly important to the development of the new fields of evolutionary psychology and sociobiology, which seek to explain aspects of our species' behaviour in terms of Darwinian adaptation. Scientists such as Leda Cosmides, John Tooby, David Buss and Steven Pinker have argued persuasively that many phenomena that occur throughout different human societies – such as aggression, co-operation, gossip, and typical male and female attitudes to sex and risk – are shared because they have evolved. These traits are found everywhere because, at least in times and places past, they helped humans to survive and thrive, ensuring that plenty of copies of the genes that influence them spread through the gene pool. Selfish genes have helped to make people who they are.

the condensed idea
Genes look selfish, but people needn't be

16 The blank slate

Karl Marx: 'All history is nothing but a constant transformation of human nature.'

Biological determinism has always had a strong intellectual rival, in the view that it is nurture, and not nature, that is primarily responsible for forming human traits. This alternative philosophy, which elevates cultural and social influences to central importance, came to dominate academia from the middle of the 20th century.

Just at the time when molecular biology was starting to unravel the secrets of DNA, genetics and evolution were relegated to bit-players by this new orthodoxy, which held that biology has forged a human mind of almost limitless malleability. Its adherents argued that to all intents and purposes, people are born as 'blank slates'.

The doctrine of the blank slate, which argues that humans share few innate character traits and instead develop them through experience and learning, is usually traced to the 17th-century philosopher John Locke – though earlier versions had been advanced by Aristotle, St Thomas Aquinas and the Islamic thinker Ibn Sina. It grew popular in the Enlightenment, fitting the mood of challenge to the authority of monarchy and aristocracy: if human capacities were not innate but learned, there was little to justify hereditary rule. For Locke, the blank slate was a statement of individual freedom.

It was later to become strongly associated with the political left. Though many early socialists had been enthusiasts for eugenics, later generations grew suspicious of genetics, from the way it was used to justify oppression of disadvantaged racial and social groups, most brutally in Nazi Germany. Liberal opinion turned decidedly against the concept of a biological human nature, which was increasingly seen as a tool with which male and bourgeois elites could rationalize their hegemony.

timeline

17th century
John Locke (1632–1704) advances blank slate theory

early 20th century
Work of B.F. Skinner (1904–90) and Franz Boas (1858–1942) popularizes social science model of human development

The social science model What came in its place was a modern formulation of the blank slate theory that had its roots in the social sciences. From psychology came Sigmund Freud's famous notion that an individual's attitudes and mental health can be explained in terms of childhood experience. To this was added the behaviourism of B.F. Skinner, which contended that humans could be conditioned into all sorts of responses by the right kind of training, much as Ivan Pavlov's celebrated dogs salivated at the sound of a bell.

From anthropology came the research of Franz Boas and Margaret Mead, whose comparative studies of different societies suggested that traditions could steer human behaviour in a multitude of directions. Mead's purported discoveries of peace-loving tribes in New Guinea and free love among Samoan women were especially influential with countercultural groups because – though founded on poor data – they challenged prevailing views about sexual mores and violence.

These ideas also suited Karl Marx's political and economic theories, which saw human nature as something that could be reshaped and directed to facilitate the revolution, and became highly attractive even to many left-

1984

Futuristic dystopias often invoke genetic determinism, but the most celebrated dystopia of them all exposes the brutal potential of the opposite philosophy. In George Orwell's *1984*, the government agent O'Brien explains to Winston Smith that his fellow dissidents will never overcome the party, because the party can mould its subjects' behaviour to suit its purposes. 'You are imagining that there is something called human nature that will be outraged by what we do, and will turn against us,' O'Brien says. 'But we create human nature. Men are infinitely malleable.'

Big Brother's apparatchik sounds distinctly like Margaret Mead: 'We are forced to conclude that human nature is almost unbelievably malleable, responding accurately and contrastingly to contrasting cultural conditions.'

1928
Margaret Mead (1901–78) publishes *Coming of Age in Samoa*

1975
E.O. Wilson (b.1929) has lectures picketed after publishing *Sociobiology – the New Synthesis*

1984
Steven Rose (b.1938), Leon Kamin (b.1928) and Richard Lewontin (b.1929) publish *Not in Our Genes*

wingers who were not Marxists. So, too, did the postmodernist mantra that behaviour and knowledge are socially constructed, and all truths are relative.

What emerged was what Leda Cosmides and John Tooby call the standard social science model of human behaviour. In this paradigm, human nature is anything but fixed or shared, but can be moulded into all sorts of configurations by the right cultural conditioning. If genetic influences are allowed at all, they are wholly secondary to those of the environment. To its supporters, this model became axiomatic to a fair society: if anything can be learned, and anybody can do the learning, then people can be taught to value equality. Social justice and morality became intertwined with the concept that little in life is laid down, or even much affected, by inherited genes.

Not in our genes Many of those who promoted this philosophy had the admirable intention of promoting individual liberty, and of fighting the injustice that unscientific genetic determinism had often been invoked to defend. It grew popular among liberal scientists, such as Stephen Jay Gould, as well as among social scientists and cultural critics. But it was also dangerously inflexible to any new scientific discoveries that might suggest human nature was genetically influenced after all. Any such evidence would threaten the very foundations of liberty and equality – so such evidence would have to be resisted, as would the sorts of research that might provide it.

The result was that scientists who advanced evidence for genetic effects on human behaviour found their positions caricatured and their politics demonized as reactionary, even fascist. E.O. Wilson, the great evolutionary theorist and conservationist, is no man of the right. Yet in the 1970s when

‘Once [social scientists] staked themselves to the lazy argument that racism, sexism, war and political inequality were logically unsound or factually incorrect because there was no such thing as human nature (as opposed to morally despicable, regardless of the details of human nature), every discovery about human nature was, by their own reasoning, tantamount to saying that racism, war and political inequality were not so bad after all.’

Steven Pinker

‘If genetic determinism is true, we will learn to live with it as well. But I reiterate my statement that no evidence exists to support it, that the crude versions of past centuries have been continually disproved, and that its continued popularity is a function of social prejudice among those who benefit most from the status quo.’

Stephen Jay Gould

he dared to suggest that human nature, like that of other animals, has a biological basis that might fruitfully be studied, his lectures were picketed and students doused him with water.

The left-wing biologists Steven Rose, Leon Kamin and Richard Lewontin responded with a book entitled *Not in Our Genes* (1984), which accused Wilson, Richard Dawkins and other sociobiologists of a crude determinism designed to legitimize the status quo. ‘Its adherents claim, first, that the details of present and past social arrangements are the inevitable manifestations of the specific action of genes,’ they said.

Such attacks were misconceived for two reasons. First, they set up a straw man. It is simply impossible to find serious biologists who believe that behaviour and social structure are ‘the inevitable manifestations of the specific action of genes’. Those who reject the social science model make a much more modest proposal – that genes, as well as culture and the environment, make a contribution to the human condition. As Dawkins wrote in a review of *Not in Our Genes*: ‘Reductionism, in the “sum of the parts” sense, is obviously daft, and is nowhere to be found in the writings of real biologists.’

What is more, cultural determinism can be just as inimical to human freedom as its genetic counterpart. It implies that instead of being prisoners of our genes, we are prisoners of our parents, teachers and societies. Those who grow up in poverty will be forever disadvantaged, while those who come from privilege will retain it. Autism can be blamed on ‘refrigerator mothers’, and adults’ relationship problems on their overprotective families. This world view is quite as bleak as the one that maintains such traits are all inherited in our genes. It certainly has little to do with social justice.

the condensed idea
Culture is important, but not all-important

17 Nature via nurture

Francis Galton: 'The phrase "nature and nurture" is a convenient jingle of words, for it separates under two distinct heads the innumerable elements of which personality is composed.'

The monster Caliban, according to his master Prospero, was 'a devil, a pure devil, on whose nature nurture can never stick'. Yet only a few decades before Shakespeare wrote *The Tempest*, St Ignatius Loyola had founded the Jesuit order, with its famous maxim: 'Give me the child until he is seven, and I will show you the man.' The debate over the relative contributions of inheritance and experience to the human condition has deep historical roots.

As we have seen, it became one of the most politically charged questions of the genetic age. On one side stood those who sought and saw genetic explanations for human psychology; on the other, those who believed it to be moulded by culture. There was not supposed to be much common ground. Sarah Blaffer Hrdy, an evolutionary psychologist, has even joked that perhaps we are genetically programmed to set nature against nurture.

The two camps, however, are not pitched so far apart as is usually assumed. Partisans of both have often caricatured one another's position, and many of their disagreements are really about emphasis. Few, if any, members of the 'nature school' are true genetic determinists who believe every human trait can be directly mapped to triplets of DNA. Equally, while strong cultural determinism is more common, most critics of genetic theories argue that the importance of genes is exaggerated rather than nonexistent. The great controversy, indeed, is now starting to give way to consensus, as

timeline

1934
Asbjørn Følling (1888–1973) identifies phenylketonuria

1953
Discovery of the double-helix structure of DNA

improved understanding of how genes actually work makes it clear that the two forces are often impossible to separate.

A genetic and environmental disease In 1934, a Norwegian doctor named Asbjørn Følling began treating two young siblings, who had seemed normal at birth but were soon diagnosed as mentally retarded. Følling conducted urine tests that revealed an excess of the amino acid phenylalanine. He had discovered the cause of the regression: an inherited disease called phenylketonuria, also known as PKU.

People with PKU have two copies of a recessive gene, which means they fail to make an enzyme called PAH. As a result, they are unable to convert phenylalanine into another amino acid, tyrosine, causing a chemical imbalance that impairs brain development. The condition, however, is treatable. If it is detected sufficiently early, infants with PKU can be put on a low-phenylalanine diet, involving no breast milk and then restrictions on meat, dairy produce, legumes and starches. This mitigates brain damage, and can allow normal development.

Which environments matter?

As few psychological attributes are entirely genetically determined, the environment must be important. But which factors matter most? You might assume that the family home is paramount, but except in cases of neglect or maltreatment this isn't generally the case.

American psychologist Judith Rich Harris has shown that children's shared home environment has very little influence over most aspects of their development: what really counts are their friends. Just as the children of immigrants pick up the accent of their peers, not their parents, so they are more likely to share their peers' social attitudes and personality traits.

Parents can teach their children skills, like playing the piano, but they can't much affect their underlying aptitude for music. And while they obviously influence their children's happiness, that won't necessarily shape their outlook on life.

2001
First drafts of human genome published

2002
Dunedin cohort study reveals genetic and environmental contributions to multiple health and behavioural effects

The cause of PKU involves both nature and nurture. Neither genotype nor diet will on its own bring about mental retardation: only the wrong combination is harmful. This discovery has helped thousands of children, as newborns are now routinely screened for the mutation so that brain damage can be prevented before it sets in.

Twin studies Many similar combination effects have emerged through the study of twins. Identical twins share all their DNA, while fraternal twins share only half – they are no more closely related on a genetic level than are ordinary siblings. Both kinds of twins, however, share a womb, a family and a cultural environment. Comparisons between the two types can thus tease out the extent to which inheritance is important.

Across a wide range of traits, including IQ, personality indicators such as extroversion and neuroticism, and even homosexuality, religiosity and political conservatism, identical twins are more similar to one another than are fraternal pairs. This indicates that genes must affect these aspects of personality.

The concordance between identical twins, however, is rarely 100 per cent – their IQ scores, for example, tend to be around 70 per cent similar, compared with around 50 per cent for non-identical pairs. By definition, inheritance therefore cannot be the only factor involved: if it were, identical twins would always turn out the same. For most human qualities, neither the extreme-nurture nor the extreme-nature hypothesis can be correct.

The Dunedin cohort study Even more striking evidence has come from a recent series of studies led by Avshalom Caspi and Terrie Moffitt. These

The genetics of aptitude

There is often an element of chicken-and-egg about the interaction of nature and nurture. Let's take sporting ability as an example. If a boy has inherited genes that give him strong fast-twitch muscles and a good lung capacity, he is quite likely to find himself a quicker sprinter than many of his peers. As a result, he's likely to enjoy sport, to attract the attention of his school's track coach, to make the 100 metres team, and to get training that adds further to his speed. He seeks out an environment that suits his genes.

Something similar probably applies to other abilities, such as intelligence and music. Genes might not influence intelligence per se, so much as create an aptitude for learning, so that a child concentrates in class and spends their free time in the library.

scientists have been following up a cohort of children born in 1972–73 in Dunedin, New Zealand, recording details of their life experiences and testing their DNA. The results have demolished the nature–nurture dichotomy.

First, Moffitt and Caspi studied a gene called MAOA, which has two variants or alleles. Boys with one allele are more likely to behave antisocially and get into trouble with the law – but only if they were also maltreated as children. When raised in well-adjusted families, those with the 'risky' allele are fine. It is not a gene 'for' criminality, and no determinism – genetic or environmental – is involved. A genetic variant must be activated by an environmental influence to do any potential harm.

The serotonin transporter gene, 5HTT, also has two alleles, and is known to be involved in mood. Moffitt and Caspi found that people with one allele were 2.5 times more likely to develop clinical depression than those with the other – but again, only under particular circumstances. The risk only affects people who also experience stressful life events such as unemployment, divorce or bereavement – and even then, it is a matter of raised risk, not determinism. When their environments are happy, their genotypes make no difference.

'The argument about intelligence has been about nature versus nurture for at least a century. We're finding that nature and nurture work together.'

Terrie Moffitt

The team has also found that one version of a gene called COMT can raise the risk of schizophrenia, if carriers also smoke cannabis as teenagers. Its most recent discovery is that children who are breastfed have higher IQs, on average, than those who are not – but only if they also have a particular genetic profile. The small minority that lacks it gets no intelligence boost.

All this goes to show the sterility of the nature–nurture debate. The question should not be which is the dominant influence, but how they work together. Nature works through nurture, and nurture through nature, to shape our personalities, aptitudes, health and behaviour.

the condensed idea
Genes and environment work together

18 Genetic diseases

Michael Rutter: 'Most ordinary people, and even many medics, still think in terms of "genes for" particular conditions. Yet genes with large effects like that are very much the exception, not the rule.'

When genes make the news, it is more often than not in the context of disease. Headlines regularly proclaim the discovery of 'Alzheimer's genes', 'breast cancer genes', even 'obesity genes'. We know that the gene for Huntington's disease lies on chromosome 4, and that the gene for sickle-cell anaemia lies on chromosome 11. Embryos can be tested for the cystic fibrosis gene or the haemophilia gene, so that only healthy ones are transferred to the womb.

You could thus be forgiven for assuming that the main function of many genes is to cause disease. Yet as the science writer Matt Ridley has pointed out, this is as misleading as it is to define the heart by heart attacks, or the pancreas by diabetes. In truth, there are no such things as 'genes for' particular diseases. The 'Huntington's gene' is not carried only by people with the devastating neurological disease. We all have it. What is different about Huntington's patients is that they carry a version with a destructive mutation. If you like, they have a 'patho-gene'.

Many of the genes that are commonly described as being 'for' this disorder or that are not even deterministic. The genes BRCA1 and BRCA2, for instance, are so closely identified with breast cancer that they are named after the disease. Women who carry mutated copies have a very high lifetime risk of breast tumours, of up to 80 per cent. But by definition, that means that at least 20 per cent of carriers will not get breast cancer. Genes like these are known as incompletely penetrant – they influence disease, but they do not inevitably cause it.

timeline

1865
Mendel presents laws of inheritance

1993
Discovery of Huntington's disease mutation

Simple and complex inheritance Some mutations, of course, do have an inevitability about them. If you inherit too many repeats of the triplet CAG in a particular gene, you will get Huntington's disease. The number of repeats can even tell you the age at which you are likely to start to experience tremors, mood swings and neurological damage, leading to death. With 40 repeats, you'll stay healthy, on average, until the age of 59, but with 50 repeats, you'll fall ill in your late 20s.

Huntington's is one of the very few examples in which absolute genetic determinism holds true. People can escape from these mutations only if science develops a treatment, or if something else kills them first. More than 200 such conditions are known, and they are in general passed on through normal Mendelian inheritance. There is a simple match between genotype and phenotype, between mutation and disease.

Some are autosomal (carried on non-sex chromosomes) and dominant, meaning that inheriting one copy is enough to cause disease: examples include Huntington's and hereditary non-polyposis colon cancer. Others, such as cystic fibrosis and sickle-cell anaemia, are autosomal recessive.

Autism

Even when medical conditions are heavily influenced by inheritance, it can still be infuriatingly difficult to trace the genes that are responsible. Autism, for example, is known from twin and family studies to be highly heritable, indicating that genes are strongly involved. Despite years of research, however, no genes that definitively predispose to this developmental disorder have yet been found.

This suggests one of two things. Either there are no 'autism genes', but the chances of developing the condition are raised or lowered by dozens or even hundreds of normal genetic variants, each of which has only a small effect by itself; or it is affected by very rare spontaneous mutations, which are unique to individuals or their families. Autism is explored further in Chapter 50.

1995
Discovery of BRCA1 and BRCA2 mutations

2001
Completion of first drafts of human genome

Only people who are 'homozygous', with two copies of the defective allele, will suffer, while 'heterozygous' carriers with just one copy experience no ill-effects. Still more are carried on the X chromosome, as with haemophilia and Duchenne muscular dystrophy, and most commonly affect boys.

Most diseases that are affected by genetics, however, are not so simple. The common conditions that are the chief causes of ill-health and death in the developed world, such as heart disease, diabetes and most cancers, are influenced by inheritance, but there is no one-to-one relationship between a particular mutation and a disease.

Sometimes, as with the BRCA genes, a defective gene has a very large effect, but not an inevitable one. More often, dozens of genes, each with small individual effects, combine to make people more susceptible to a disease. On their own, such genetic variants are virtually harmless. Put together, they explain why some families struggle with high blood pressure, while others tend to develop cancer.

Why do disease genes survive? As patho-genes like the ones that cause Huntington's and cystic fibrosis are so harmful, you might expect them to have been weeded out by evolution. Natural selection is brutal towards alleles that confer even the slightest survival disadvantage, and these have catastrophic effects. How have they managed to hang on to their places in the human gene pool?

Sometimes, the answer is pure bad luck. A spontaneous mutation in the egg or sperm from which a person is conceived can occasionally prove catastrophic, if it occurs in a vital place. Diseases caused by extra genetic repeats, such as Huntington's and fragile X syndrome (which can cause mental impairment), are especially likely to be triggered in this way. It often takes only a small error to transform an acceptable number of repeats into a damaging one.

> **'We all have the Wolff–Hirschhorn gene, except, ironically, people who have Wolff–Hirschhorn syndrome.'**
>
> **Matt Ridley**

Other deleterious mutations survive because they cause damage only late in life, long after the carrier has had time to breed and raise a family. Many genetically influenced cancers, and Huntington's again, are good examples, as most patients live well into their 50s before symptoms strike. In such circumstances, natural selection does not apply. People with these defects have just as many offspring as those without them.

Cancer and diabetes: another trade-off?

Sickle-cell anaemia is not the only disease in which a genetic trade-off is involved. Recent research has suggested that a similar effect may exist for type 2 (adult-onset) diabetes and some cancers, with the discovery of certain genetic variants that seem to raise people's risk of one disease while lowering their chances of developing the other.

What may be happening is that these genes affect the rate at which cells divide. Variants that promote division may be protective against diabetes, as they encourage the regeneration of beta cells in the pancreas that make insulin, but they may also make the unchecked cell growth of cancer more likely. Variants that slow down the cell cycle may work the other way around.

For recessive genetic disorders, another factor can be involved. Often, these have flourished because people who carry just one copy of a mutated gene have some kind of advantage. A single copy of the defect that causes sickle-cell anaemia, for example, confers a degree of resistance to malaria. The evolutionary benefits of being heterozygous can outweigh the costs of conceiving some homozygous children with a crippling illness. The sickle-cell mutation is most common in regions where malaria is endemic, and where the genetic trade-off is worthwhile.

With complex conditions such as heart disease, to which many genes contribute, the picture is different again. The variants that add slightly to risk are not best thought of as disease genes at all. These are common variations, with multiple effects – in the jargon, they are pleiotropic, from the Greek for 'many influences'. These influences can be positive as well as negative, which explains why they have spread so widely through the gene pool.

Genes are not for disease, and even rogue genes do not underlie most widespread diseases. Rather, these are influenced by the action of perfectly normal genes, working in concert with the environment.

the condensed idea
Genes are not 'for' disease

19 Gene hunting

Mark McCarthy, University of Oxford : 'We are mostly finding that for any given disease there are zero or at best one or two genes with large effects. Then there is a sprinkling of genes, perhaps five to ten, with modest effects of 10 to 20 per cent, and there may be many hundreds with even smaller effects.'

In the late 1970s, Nancy Wexler, the daughter of an American doctor, set out in search of the genetic mutation that causes Huntington's disease. Wexler's mother and uncles had the condition, and she knew there was a 50 per cent chance she had inherited it. Finding the defect, she reasoned, would allow people in her position to discover whether they had been given a genetic death sentence. It might also lead to a treatment.

On learning of an extended family with a high incidence of Huntington's in Venezuela, Wexler travelled to Lake Maracaibo in 1979 to collect blood from more than 500 people. She then sent the samples for genetic analysis by her collaborator, Jim Gusella. His team began to compare DNA taken from people with and without Huntington's, and by 1983 he had narrowed down the search to the short arm of chromosome 4. It took another decade, however, to identify an actual gene, which makes a protein called Huntingtin.

The discovery, in 1993, was one of the great early successes of disease genetics, but it was the result of an extremely laborious process. The project took 14 years to deliver – and though it has led to a test (which Wexler decided not to take), it has yet to produce a therapy.

The Huntington's mutation, too, was among the genome's low-hanging fruit. It has a catastrophic effect, and it is autosomal dominant, inherited in simple Mendelian fashion. These factors meant it would be one of the

timeline

1976
Nancy Wexler (b.1945) starts search for Huntington's mutation

1993
Wexler's team identifies the Huntingtin mutation on chromosome 4

easier genes to find. Other genes that influence disease in subtler fashion, however, were going to be very difficult to pin down.

Linkage analysis The Huntingtin gene was identified using a technique called linkage analysis, which was until recently the most effective way of detecting how genetic variations influence disease. It relies on the way genes that rest close together on chromosomes tend to be inherited together, because of the recombination effect discussed in Chapter 6.

> **‘We have now entered a new era of large-scale genetics unthinkable even a few years ago.’**
> **Peter Donnelly**

First, scientists select a number of single nucleotide polymorphisms (SNPs) – DNA sequences that are known to vary in spelling by one letter – as markers spaced at appropriate intervals through the genetic code. The next step is to search for these markers in people from families in which an inherited disease like Huntington's occurs. If a marker is always found in people with the disease, but not among those who are healthy, it must lie close to the mutation responsible, which can then be identified and sequenced. As family members share so much of their DNA, it is usually necessary to look only at a couple of hundred markers, in a few dozen people, to get results.

This technique, however, is readily applicable only to fairly rare disorders caused by mutations with large effects, as with Huntington's or BRCA1 (see Chapter 18). To find subtler influences on disease, many more people have to be screened. The numbers required make it essential to look beyond families, to less closely related people who share less of their DNA. That, in turn, means that hundreds of thousands of genetic markers have to be scanned, to get a statistical relationship strong enough to reveal a gene. Until recently, that was so expensive and time-consuming as to be functionally impossible.

Genome-wide association It has now become practical with the advent of two new tools that have transformed disease genetics. The first is the micro-array or ‘gene chip’ (see box), which can screen a person's DNA

2001
Completion of first drafts of human genome

2005
Completion of HapMap makes genome-wide association a viable research tool

2007
First wave of genome-wide association studies published

for a million genetic variations at once. The second is the HapMap, a chart completed in 2005 that shows which segments of the genome, called haplotypes, tend to be inherited together.

The new technique, called genome-wide association, starts with the HapMap, from which scientists select 500,000 SNPs as markers for every haplotype block. Gene chips are then used to look for these markers among thousands of people with a particular disease – say type 2 diabetes – and a similar number of healthy controls. Any markers that are significantly more common in either group are then investigated further, to pinpoint parts of the genome that are associated with a higher or lower risk.

Gene chips

Research projects like the CCC would not have been possible without the development of DNA micro-arrays or 'gene chips'. These hold a collection of up to a million microscopic spots of DNA, each in the configuration of a particular SNP. When the DNA that you wish to test is exposed to this chip, any sequences that are present will bind to the corresponding spot. They can be used to screen for hundreds of thousands of genetic markers at once, revealing which SNPs are carried by the person who is being tested.

The beauty of this method is that it can reveal completely unexpected results. If a variant raises the risk of a condition by more than about 20 per cent, genome-wide association will find it, even if this effect had never been suspected. A variant in a gene called FTO, for example, causes fused toes in mice. In 2007, one of the first large genome-wide association studies, by the Wellcome Trust Case Control Consortium (CCC), found that in people, it predisposes slightly to obesity.

Early in 2007, science knew of hardly any common genetic variants that influenced disease. By the spring of 2008, the tally stood at more than 100, as genome-wide association studies began to bear fruit. The CCC has found genes linked to heart disease, rheumatoid arthritis, Crohn's disease, bipolar disorder and both kinds of diabetes, as well as obesity and height. Other teams have found new variants that affect breast cancer, prostate cancer, heart attacks and multiple sclerosis. Exciting new data are being published all the time. Even cautious voices are speaking openly of a step-change in humanity's ability to read and understand the genome.

Each of these variants has a small effect on its own, raising risk by between 10 and 70 per cent. When considered alongside other variants, however, their combined effects can be large. They are also extremely common – those identified by the CCC are carried by between 5 and 40 per cent of the Caucasian population. As the diseases they influence are common, they clearly affect hundreds of millions of lives.

Genetics is suddenly being taken to a different level. It was once a science limited to finding mutations with devastating effects, but for very few people. It is now tracing variants with a more limited impact, but on common medical conditions. You might call it the democratization of the genome.

The 1,000 genomes project

One of the next stages in the hunt for genes that affect our health is an international effort to map the entire genomes of more than 1,000 people, which has been made affordable by new sequencing technology. This should allow scientists to find and catalogue every single genetic variant that is carried by at least one person in 100. It will effectively work as an index to the genome. When a marker SNP suggests that a part of the genome is linked to a disease, geneticists will immediately be able to look up all the reasonably common variants in its chromosomal neighbourhood, to investigate which, if any, are responsible for the effect.

the condensed idea
Common genetic variants can affect disease

20 Cancer

Mike Stratton, Cancer Genome Project leader, 2000: 'It would surprise me enormously if in 20 years the treatment of cancer had not been transformed.'

Despite the fact that most common diseases are the result of complex interactions between inheritance and our surroundings, the products of nature via nurture, there is one that always features genetics at its core. Rather, it is not one disease, but a group of more than 200 – the cancers. Brain and breast tumours, carcinomas of the lung and liver, melanomas of the skin and leukaemias of the blood share a common characteristic. They are ultimately diseases of our genes.

That may come as something of a surprise, given that cancer is often thought of as an environmental disease. Whether it is sunbeds and melanoma, the human papilloma virus and cervical cancer, asbestos and mesothelioma, or smoking and any cancer you choose, there is overwhelming evidence that environmental influences can contribute, often decisively, to the formation of tumours. All these carcinogens that can seriously damage your health, however, do so in essentially the same way. They damage DNA.

Cancer is the result of genetic failure. Every time a cell divides, it must successfully copy its DNA. It is estimated that 100 million million cell divisions take place during an average human lifetime. Every one has the potential to introduce an error into a daughter cell's genetic code that can turn it cancerous.

In healthy tissue, cell division is a controlled process, ordered by genetic signals that ensure it happens only when it is supposed to. Cancer develops when it starts to run out of control. In every case, the trigger is a copying mistake during cell division, often at a single DNA letter. Many mistakes

timeline

1953
Discovery of double-helix structure of DNA

1971
Richard Nixon declares 'War on Cancer'

of this sort are harmless, doing nothing to alter the genome's function, but when mutations strike in the wrong place, the results can be disastrous.

Oncogenes and tumour suppressors The genetic errors that start cancer can be inherited or acquired from exposure to carcinogenic chemicals or radiation. To launch the destructive career of a tumour, though, they need to affect two broad categories of gene. The first class are the oncogenes – genes which, when defective, give cells new properties that turn them malignant. The second are the tumour suppressors – the genome's policemen, whose job is to spot oncogene mutations and tell malignant cells to kill themselves.

Most cells that acquire oncogene mutations are shut down by their tumour suppressors, and commit suicide by a process called apoptosis. A cell with mutations in both kinds of gene, however, can escape this programmed death and become cancerous, though sequential damage to many different

Telomeres

Another genetic clue to cancer comes from stretches of repetitive DNA at the end of each chromosome called telomeres, which protect against the loss of genetic information. Without them, some important genes would be disrupted every time a cell divides, because DNA cannot normally copy itself all the way to the end of the chromosome. The telomeres absorb this damage, shortening a little with each cell division, and when they are lost completely, the cell usually dies. Telomere loss is one of the main causes of ageing.

One of the reasons why cancer cells grow out of control is that most can copy their telomeres, because of mutations that allow them to make an enzyme called telomerase. This helps them to divide unchecked, but it also suggests a medical plan of attack. Several telomerase-inhibitor drugs have begun clinical trials.

1986
Renato Dulbecco proposes sequencing the human genome to improve understanding of cancer

2003
Cancer Genome Project links BRAF gene to malignant melanoma

2008
Launch of International Cancer Genome Consortium

genes is usually required. The cell will divide unchecked, passing its mutant genetic legacy to its progeny, which proliferate to create rogue tissue that can eventually metastasize through the body, damage organs, and kill.

Many of the oncogenes that drive cancer are implicated in tumours found in very different parts of the body. Mutations in the BRAF gene, for instance, are common in both malignant melanomas, in which they are often caused by ultraviolet light, and in colon cancer. The same tumour suppressors are often damaged too – the p53 gene is mutated in almost half of all human cancers. Most inherited mutations that contribute to cancer affect tumour suppressors as well – both BRCA1 and BRCA2 have this role. These defects greatly raise the lifetime risk of cancer by reducing by one the number of genetic hits a cell must take to set it on the path to malignancy.

Genetic therapy To treat cancer, it is necessary to root out the genetically abnormal cells that cause it, either by killing them with drugs or radiation, or removing them with surgery. All these methods can be pretty brutal: operations such as mastectomies can be disfiguring, while chemotherapy and radiotherapy respectively poison and burn healthy tissue as well as the tumours they are designed to cure. Their side-effects are legion.

‘I imagine that in the future, machines that read the genetic signatures of patients’ cancers will be more important than their oncologists.’

Richard Marais, Institute of Cancer Research

These blunt instruments, however, are now being supplemented with smarter weapons, and genetics provides the guidance system. If it is possible to characterize the precise genetic mutations that are driving a particular cancer, it is also possible to target these with drugs. A prime example is Herceptin, a drug prescribed to women whose breast tumours have mutations in the gene for a receptor called HER-2. The drug binds to this receptor, killing the cancer. It can halve the relapse rate – but only among those patients whose cancers are genetically susceptible to it. In others, it has no effect. Had it been tested in the general population, rather than in a targeted group, it would never have made it through clinical trials.

This is the future of cancer treatment, which a project called the International Cancer Genome Consortium should help to realize. This $1 billion initiative aims to identify all the mutations that drive 50 common types of cancer, so that doctors can pinpoint the precise genetic factors that are responsible for the growth and spread of their patients' tumours. Cancers could then be treated not so much according to where they occur in the body, but on the basis of the genetic makeup of their rogue cells. We may soon think not of bowel or stomach cancers, but of BRAF-positive or p53-positive tumours.

Mike Stratton, of the Wellcome Trust Sanger Institute, a leader of the consortium, is already seeking to develop therapeutic strategies based on this approach. His team is currently investigating how 1,000 cancer cell lines, each with known mutations, respond to 400 different drugs. The goal is to determine whether some of these agents are effective against tumours with a particular DNA profile, but not against others.

Another benefit of cancer genomics should be to make chemotherapy kinder, through drugs that home in on DNA targets that are found in cancer cells, but not in healthy tissue. It may also be possible to avoid damage to a patient's reproductive cells: these are particularly vulnerable to existing cancer therapies, which often cause sterility as a consequence.

The cancer paradox While both life expectancy and its quality have improved significantly in Western countries in the past century, cancer rates are going up. Between 1979 and 2003, the UK incidence rose by 8 per cent in men, and by 26 per cent in women. This is sometimes blamed on pollution or other environmental factors, but its principal cause is actually the success of modern medicine.

Antibiotics, sanitation, better nutrition and other improvements in public health mean that fewer people are dying young of infectious diseases – but longer lives allow more DNA damage to accumulate, to the point at which tumours can form. The genetic nature of this disease explains the apparent paradox of medicine. As it defeats other foes, more of us will live long enough to develop cancer. The challenge, which genetics will help to meet, is to turn it from a fatal disease into a chronic one.

the condensed idea
Cancer is a disease of the genes

21 Superbugs

Jared Diamond: 'Diseases represent evolution in progress, and microbes adapt by natural selection to new hosts and vectors.'

Not every disease has a genetic origin quite as obvious as cancer, Huntington's or even diabetes. Yet as the Nobel laureate Paul Berg has said, all disease is genetic to some extent. Infectious diseases such as HIV/Aids, tuberculosis and flu are not caused by DNA damage, as are tumours, or by major Mendelian mutations, like cystic fibrosis. But the genes of both pathogens and their human hosts are pivotal to the way viruses, bacteria and parasites make us ill.

The T-cells, lymphocytes and antibodies of the immune system, which protect our bodies against germs, are all affected by our genetic makeup, and slight variations can make us more or less susceptible to certain diseases. People with type O blood – a genetically determined trait – are less vulnerable to malaria, and those with other genotypes are less susceptible to HIV.

Genes also control how pathogens attack, and how they fend off the immune system and the drugs and vaccines with which medicine supports it. They explain why some strains of flu confine us to bed for a day or two, while others kill millions in months. They explain how new diseases emerge to sweep through populations, and how medicines that once worked have gradually become useless. And genetic insights into infections can highlight ways in which they can be stopped.

Evolution and disease When Christopher Columbus reached the New World in 1492, it is estimated that perhaps 50 million people lived on the continents of North and South America. Yet by the middle of the

15th & 16th centuries
Germs taken from Europe to the Americas devastate native populations

1928
Discovery of penicillin

17th century, this indigenous population had collapsed to between 6 and 8 million. Some were certainly victims of genocide by the colonial invaders. The most fearsome killers, though, were not the Spanish conquistadors, but diseases that hitched a ride on their transatlantic voyages.

For centuries, the peoples of the Old World had lived with smallpox and measles, typhus and yellow fever. As a result, they had evolved a measure of resistance: natural selection had favoured genes that improved the chances of surviving these infections. Native Americans, by contrast, were immunologically naïve. Their smallpox-free environments had not promoted the spread of random mutations that build resistance. When the virus arrived, there was nothing to hold it in check. As scientist Jared Diamond has recounted in *Guns, Germs and Steel* (1998), Spanish diseases were at least as important to the rapid conquest of the continent as Spanish technology.

A similar process explains how new infectious diseases have repeatedly jumped the species barrier from animals to humans. HIV, the virus that causes Aids, is thought originally to have been a chimpanzee infection that crossed to humans during the 1960s or 1970s, probably when a bushmeat hunter was bitten. Though harmless in chimps, people lacked genetic defences. The virus soon acquired further mutations that allowed it to spread from person to person, causing a pandemic that kills at least 2.5 million each year.

Dodging our defences Over time, some people will probably evolve resistance to HIV, just as some have developed genetic defences against smallpox and malaria. The long human life span, however, means it will take centuries for such traits to emerge by mutation, and then to spread widely through the human gene pool. Pathogens do not have this handicap: the phenomenal speed with which bacteria and viruses reproduce gives them a huge advantage over their hosts. Put simply, they can evolve more quickly than we can, to dodge the weapons with which we fight back.

1961
First identification of MRSA

2001
MRSA genome sequenced

In the middle part of the 20th century, the advent of antibiotics brought about a revolution in infectious disease control. Drugs such as penicillin and streptomycin meant that even killers such as tuberculosis and meningitis could be treated successfully more often than not. By the 1970s, many doctors spoke openly of the defeat of bacterial disease. Antibiotics became so commonplace that they are sometimes thought of as synonymous with drugs – even today, many patients with viral illnesses are disappointed when doctors will not prescribe them.

Bacteria, however, multiply so rapidly that their genomes rarely stand still for long. Each of the billions of cell divisions that a colony will experience each day presents an opportunity for mutation – and some of these mutations will confer a degree of antibiotic resistance. Natural selection means that if an antibiotic is then used in treatment, a few bacteria will survive, and then divide to seed a new colony with resistant progeny. Resistance can also spread in another way, as bacteria donate immunity genes to their neighbours by exchanging them in portable DNA parcels known as plasmids.

Thus are superbugs born. Most strains of MRSA, which stands for methicillin-resistant *Staphylococcus aureus*, are resistant to the entire penicillin family of antibiotics. Infections with this bacterium, once considered easily treatable, are now implicated in around 1,600 UK deaths each year. Tuberculosis that is immune to multiple antibiotics infects 500,000 people annually worldwide. Resistance is not confined to bacteria, either – viruses such as HIV, and parasites such as *Plasmodium falciparum*, which causes malaria, have acquired drug immunity as well.

Genetic medicine Humanity may lack the capacity to evolve as quickly as its microscopic enemies, but it has another weapon at its disposal. By studying pathogen genomes, new medicines can be designed from a position of strength. The discovery that HIV requires an enzyme called reverse transcriptase to reproduce, for example, led to the development of inhibitor drugs such as AZT, which can prevent the development of full-blown Aids for decades.

Influenza genetics has brought us neuraminidase inhibitors – drugs such as Tamiflu that interfere with a key protein that the virus needs to enter cells. These have become the frontline of world defences against a future pandemic. The genomes of the agents that cause malaria, TB, chlamydia, plague, MRSA and typhoid have been sequenced, which will allow scientists to find essential genes that might be targeted with new drugs. It is even becoming possible to identify genes that cause antibiotic resistance, which could be inhibited to restore the effectiveness of these once-powerful drugs. The pathogens' genetic advantage may not last for long.

The evolution of virulence

New pathogens are often extremely virulent, with a high death rate, because their naïve hosts have little resistance. Over time, however, they often decline in severity, and not just because evolution gradually helps the human body to fight back. Extreme lethality can be bad for a germ's adaptive fitness, too.

If a virus or bacterium kills its host too quickly, before it has a chance to infect a new one, it will also die, along with all its progeny. Natural selection can thus favour strains that cause less damage to the organisms they inhabit, as these may be more likely to spread.

This may explain why many diseases lose their virulence over time. Syphilis, for example, had a high death rate when it first emerged in Europe in the 16th century, probably imported from the New World. But while still a serious illness today, it is not normally life-threatening. New flu strains tend to follow the same path. H5N1 bird flu is currently extremely lethal, killing more than 60 per cent of people it has infected so far, but scientists predict this death rate will fall substantially if it mutates to pass easily from person to person.

This trend, however, is not inevitable. If a germ accelerates death through symptoms that help it to spread, such as sneezing, haemorrhage or diarrhoea, it is not inconvenienced by its host's demise, and can remain highly lethal.

the condensed idea
All disease is genetic

22 Behavioural genetics

Nuffield Council on Bioethics: 'It would be unwise to assume that genetics will not be able to assist in determining degrees of blame, even if the "all-or-nothing" question of responsibility is not affected by genetic factors themselves.'

Certain behaviours and personality traits are well known to run in families. People with religious parents are more likely to be churchgoers, while those who grew up in left-wing households are more likely to vote that way when they reach 18. We all know people, too, whose character quirks remind us of their close relatives – nervous daughters of nervous mothers, and fathers and sons who share a passion for fishing or fast cars.

Folk wisdom tends to attribute all this to upbringing: to the way a child's outlook on life is moulded by that of its parents, whether through deliberate indoctrination or passive exposure to their tastes. That conclusion, however, is too simple. Children of course share a home environment with their mothers and fathers, which can greatly affect personal development, but that is not all they share. They also inherit half their DNA from each parent, and the science of behavioural genetics has shown that this can be equally important, if not more so.

Natural experiments The relative contributions of nature and nurture are fiendishly hard to separate when studying families, as either could account for shared traits, from spirituality to spitefulness. As it is unethical to separate children from their parents in controlled experiments, such research must rely on natural experiments instead.

timeline

late 19th century
Francis Galton studies heritable basis of behaviour

1953
Discovery of double-helix structure of DNA

As we saw in Chapter 17, identical twins share both a home environment and all their DNA, while fraternal sets share the same home but only half their genes. Comparisons between the two are therefore illuminating; for traits affected by genetics, identical pairs will look more alike. Adoption studies are also useful. For characteristics that are strongly heritable, adopted children should conform more closely to their birth families than to their adoptive ones.

Such studies have shown that genetics does not just affect physical attributes such as height and obesity. Many aspects of mental, psychological and personal development are at least partially heritable, too. The list includes intelligence, antisocial behaviour, risk-taking, religiosity, political views, and all the 'big five' personality traits – neuroticism, introversion/extroversion, agreeableness, conscientiousness, and openness to experience. There is even evidence to suggest that a woman's ability to have an orgasm may be influenced by her genes.

Heritability These effects can be quantified, using statistical techniques to calculate heritability quotients. These are expressed as a percentage or decimal, and they are easily misunderstood. When behavioural geneticists say that a trait, say thrill-seeking, is 60 per cent heritable, it does not mean that any given person can attribute 60 per cent of his aptitude for bungee-jumping to his genes. Neither does it mean that of 100 people who like extreme sports, 60 have inherited this passion while 40 have learned it. The true

Height

One pitfall of behavioural genetics is well illustrated by a non-behavioural trait in which genes are certainly involved – height. It is estimated that about 90 per cent of the differences between individuals' heights reflect genetic variation, and 20 genes that are involved have been identified. Though environmental effects like nutrition matter, the influence of genetics is strong.

Nobody of sound mind, however, would suggest that height should be assessed by genetic tests. You will always get more accurate answers by measuring people. The same is true of all sorts of heritable characteristics, such as personality, intelligence or aggression. When a phenotype can be reliably evaluated, the genotype that contributed to it is often irrelevant in the real world.

1970s
Sociobiology movement suggests evolutionary influences on human behaviour

late 20th century
Twin studies demonstrate heritable influences on multiple personality and behavioural traits

1995
Stephen Mobley appeals murder conviction on grounds of his genetic profile

implication is much subtler: it is that 60 per cent of the differences we find between different people's attitudes to risk are down to inherited variation.

To call a trait heritable is thus meaningful only at a population level; it says nothing about precisely how genetics has affected a particular person. In some, genes will be the most important factor, while in others it will be formative experiences. Heritability quotients reflect an average. Unless the value is zero (as for the language you speak) or one (as for Huntington's disease), both nature and nurture are always involved.

It is a misconception that findings of heritability imply genetic determinism. If anything, the reverse is true: most heritability quotients for behaviour and personality hover between 0.3 and 0.7, leaving a large role for environmental influences.

Ethical conundrums Most of the time, this sort of research is benign. If we learn how far genetics is involved in reading disabilities, or antisocial behaviour, it may be possible to identify the genes – or the environmental factors – that play a part, and develop drugs or social programmes to intervene. But knowledge about genetic effects on behaviour can also open more difficult ethical territory.

Are twins good models?

Twin studies are the backbone of behavioural genetics, but their value has been called into question. Critics suggest that twins may differ from singletons, and thus may not be representative of society as a whole. Parents might also treat identical twins more similarly than they do fraternal pairs.

Twin researchers largely dismiss these issues as serious problems. There is little evidence that twins are much different from non-twins. And when parents mistakenly believe that identical twins are non-identical, they are still more similar than fraternal sets.

In 1991, Stephen Mobley robbed a branch of Domino's Pizza in Oakwood, Georgia, and shot the manager, John Collins, dead. He was convicted of murder and sentenced to death, but his lawyers then appealed on innovative grounds. Their client came from a long line of violent criminals, and carried a genetic mutation that had been linked to similar behaviour in a Dutch family. Mobley's genes made him do it, they argued, and his sentence should be commuted as a result.

The appeal was thrown out, and Mobley was executed in 2005. Most scientists think his claim was specious, as the link between his mutation and violence is far from robust. If a number of genes are reliably shown to influence aggression or psychopathy, however, then future cases may not be quite so clear-cut.

Genetic tests are unlikely to provide a defence: genes may predispose people to patterns of behaviour, but they do not inevitably cause them. Some people argue, though, that such information could be considered in mitigation, in similar fashion to psychiatric illness. Britain's Nuffield Council on Bioethics recently suggested that genetics could helpfully 'assist in determining degrees of blame'.

Other possibilities are more sinister. Genetic profiling might be used to identify those whose genes make them more likely to become offenders, in the manner of the movie *Minority Report*. Similar techniques might be employed in schools, to select genetically gifted pupils for special tuition, or to assess job applicants for inherited aptitude for particular roles.

Applications such as these, however, would rest on a misunderstanding: behavioural genetics is probabilistic, not deterministic, and applies to populations, not individuals. To use its findings to prejudge people would therefore be a gross infringement of liberty. The way people behave is the result of a complex interaction between their genes and their experiences, and the balance may differ for every one of us. Individual capacities are best assessed by considering people as they are, not by trying to predict what they genes say they should be.

'Behavioral genetics cannot deal with highly complex behaviors, and certainly not generic ones like good and evil. There is no data of genes that predispose toward good or evil, and any such data would be so weak as to apply [only] to a minority of cases.'

Philip Zimbardo, Stanford University psychologist

the condensed idea
Heritable does not mean determined

23 Intelligence

Robert Plomin: 'This [IGF2R] is *not* a genius gene. It is not even *the* gene for general intelligence; at most it is one of many.'

Everybody knows that intelligence is not a uniform concept. There are mathematical geniuses who struggle to express themselves in fluent language, and literary scholars who are confused by all but the simplest sums. Apparently clever people often seem deficient in common sense, and an ability to figure out what is wrong with a faulty car engine does not always correlate with a large vocabulary.

Yet despite this variety of intellectual talents, most people accept the notion of a broad, cross-cutting intelligence. As long ago as 1904, Charles Spearman, an English psychologist, noted that schoolchildren's grades in different subjects tend to cluster: a pupil who scores highly in maths will usually be near the top of the English class, too. Spearman attributed this to general intelligence, or 'g'.

Spearman's findings are supported by IQ tests. Though these can measure different intellectual skills, such as speed of thought and verbal, numerical and spatial reasoning, an individual's results in all these areas tend to correlate. Despite concerns over the accuracy of IQ testing, at least some of the differences between people's mental abilities seem to be explained by g.

It also seems to be highly heritable. Twin and adoption studies indicate that a significant amount of variation in g – between 50 and 70 per cent – can be attributed to genetics. This shouldn't be surprising: like all organs, the brain's development is influenced by DNA. Yet perhaps because early assessments of intelligence were led by eugenicists like Francis Galton, the link remains controversial. Many social scientists still refuse to accept it.

late 19th century
Francis Galton studies inheritance of intelligence

1904
Charles Spearman (1863–1945) proposes idea of general intelligence, or g

Intelligence genes Though behavioural genetics shows that intelligence is heritable, it says nothing about which genes are involved. To identify them, scientists must use the comparative techniques described in Chapter 19, which are usually employed to investigate disease.

Every year, academically gifted American children, whose IQ scores average around 160, are chosen to attend a summer camp in Iowa. Robert Plomin, of the Institute of Psychiatry in London, realized that this group would be a wonderful resource for genetic research, and obtained permission to test the DNA of 50 participants. He then compared these samples with DNA from 50 children of similar age and social background who were not enrolled in the camp.

Of more than 1,800 genetic markers that he scanned, one stood out: a gene on chromosome 6 called IGF2R. A particular variation was significantly more common among the highly intelligent pupils. Perhaps IGF2R was a gene for intelligence?

That is certainly what media reports began to suggest. Plomin himself, however, was more circumspect. He pointed out that should the results be

Changing heritability

A common assumption about genetic effects on behavioural traits such as intelligence is that their influence will decline with age, as education and life experiences become more important. In fact, the reverse is true. There is a wealth of evidence that genes become more, not less, significant to our personalities with age.

How could this be so? It seems that when we are children, we are heavily affected by our family and school environments, and act and achieve accordingly. But as we grow older, we gain progressive freedom to act and achieve as individual nature and temperament dictate. We can and do throw off social pressures created by others, when we find they do not suit us.

1980s
Twin and adoption studies suggest genetic influence on IQ

1984
James Flynn (b.1934) identifies rising trends in IQ

1998
Discovery of possible link between IQ and IGF2R gene

replicated (which has not happened), this gene would be but one of many that affect intelligence. At least half of the 21,500 or so human genes are expressed in the brain, and any of these could plausibly affect mental development. Any effect of IGF2R would be a tiny element of the whole picture, accounting for a small fraction of genetic influence on g. It is not a gene for intelligence, but one of dozens of candidates, thousands even, that might each make a subtle difference.

Plomin's recent research illustrates the point. A study of 7,000 seven-year-olds has linked six genes to IQ – but each has an effect so tiny as to be barely measurable. The most powerful accounts for just 0.4 per cent of variation in intelligence, and even when all six are combined, they are responsible for no more than 1 per cent. These, too, are probably the genes that exert the strongest influence on IQ: if there were others with major consequences, they would already have been found.

The Flynn effect IQ, of course, is not a perfect measure of intelligence. Early tests were culturally specific, producing low scores for certain social or ethnic groups because they lacked the general knowledge needed to complete them. Modern versions largely avoid this pitfall, but their results are problematic for another reason: at least in developed countries, average scores keep going up.

This phenomenon is known as the Flynn effect, after New Zealand academic James Flynn, who first identified it. It is often used to question the claim that genes make an important contribution to intelligence. If it were genetically determined, critics argue, IQ scores should be stable. Either the tests are unreliable, in which case the whole field of inquiry is flawed, or intelligence must be a product of environmental factors, which can change much more rapidly than genes.

IQ tests are prone to error, but they do have value. They predict academic performance, independently of family background, and they provide at least a crude measure of intelligence. The Flynn effect, however, need not refute the genetics hypothesis. Not even the most bullish behavioural geneticist would say that intelligence is unaffected by nurture: its heritability of between 0.5 and 0.7 by definition means that the environment is involved. And even when traits have a large genetic component, environmental differences can still have a big impact.

Nobody would dispute that height is influenced by genes: it is among the most heritable of all human traits, with 90 per cent of variation attributable to DNA. Yet in developed countries, average heights increased by about 1 cm per decade between 1920 and 1970. This is entirely the result of environmental factors, such as better nutrition and healthcare: the timescale is too short for genetic evolution to be involved. Even when genetic effects are powerful, there is still room for substantial environmental variation.

The Flynn effect suggests that something similar is happening with intelligence. Indeed, as intelligence is less heritable than height, environmental influence will be greater. Improved diet and education, the increasing importance of technology, and changes in family structure and the labour market, could all be affecting mental development. But that does not preclude a strong influence for genes.

ASPM

A gene called ASPM is known to be involved in brain growth. Its size is related to the number of neurons in the adult brain of different species – it is longer in humans than in mice, and longer in mice than in fruit flies. When it is defective, the result is microcephaly, a disorder that stunts brain growth.

Bruce Lahn, of the University of Chicago, has discovered that a new human ASPM allele emerged around 5,800 years ago, and has since spread rapidly – a strong hint that it provides a selective advantage. As the variant started to proliferate at about the same time that humanity first developed agriculture, settled in cities and used written language, it has been suggested that this advantage might be related to intelligence.

So far, the evidence does not support this: ASPM profiles do not seem to influence IQ scores, though there are tentative links to proficiency with tonal languages like Chinese. But it is entirely plausible that other recently evolved genotypes may indeed have an effect.

the condensed idea
Genes affect intelligence

24 Race

Richard Lewontin: 'Human racial classification is of no social value and is positively destructive of social and human relations. Since such racial classification is now seen to be of virtually no genetic or taxonomic significance either, no justification can be offered for its continuance.'

While promoting a new book in the autumn of 2007, James Watson gave a newspaper interview in which he pronounced himself 'inherently gloomy about the prospects of Africa'. Development policies, he said, were based on the idea that Africans are as intelligent as Westerners, 'whereas all the testing says not really'. While he hoped that everyone was equal, 'people who have to deal with black employees find this not true'.

Watson's suggestion that there are inherited racial differences in intelligence, which disadvantage black people, unleashed fierce controversy. Scientists lined up to attack his views as founded on prejudice, not evidence. Invitations to speak were rescinded, he was suspended from his academic post, and he retired soon after the furore.

Why, though, did these remarks prove so inflammatory? Intelligence is known to be heritable, and it is not impossible that ethnic groups have evolved different average capacities. Was the great geneticist unfairly vilified for airing opinions that were politically incorrect, but scientifically valid?

Race and intelligence Watson was not the first scholar to suggest that races might have different innate abilities. The view was widely held in the 19th century: Charles Darwin wrote, in *The Descent of Man*, that mental characteristics of human races are 'very distinct; chiefly as it would appear in their emotional, but partly in their intellectual faculties'. In the

timeline

1871
Charles Darwin's *The Descent of Man* suggests ethnic differences in behaviour

1972
Richard Lewontin argues that race is biologically meaningless

1960s and 1970s, IQ testing revealed that ethnic groups did not perform in uniform fashion. In the United States, African-Americans consistently scored below Caucasians, while people of East Asian and Ashkenazi Jewish origin, on average, performed better than both.

The notion that this variation might be innate was most famously advanced in *The Bell Curve*, a 1994 book by Richard Herrnstein and Charles Murray. Researchers such as Richard Lynn and Philippe Rushton have gone further, to argue that natural differences in IQ may explain global inequalities – the point that Watson made about Africa. Henry Harpending, of the University of Utah, has proposed that Ashkenazi intelligence may be linked to the Jews' history of persecution, and to their traditional roles in trade and money-lending. Such selective pressures might have favoured genes that promote mental agility, which would have spread rapidly in closed Jewish communities whose members rarely married non-Jews.

Athletic ability

The Olympic 100 metres was last won by a white man, Allan Wells, in 1980, when a US boycott meant that the fastest black sprinters did not compete. Black athletes are also over-represented in top-level soccer, American football and basketball teams. This has fed a widespread perception that black people have a genetic advantage in some sports, particularly those that reward pace and power.

This could be correct: there are genes, such as ACTN3, which affect the fast-twitch muscle fibres that produce explosive speed, though there is little evidence that they vary with race. Black athletic achievement could also reflect social circumstances or cultural traditions, which steer athletically talented people of different ethnic origins towards different sports.

Sport also illustrates the limitations of traditional racial categories. As Jon Entine points out in his book *Taboo* (2000), sprinting events are usually won by athletes of West African descent, while East and North Africans are outstanding over middle and long distances. Ethnicity and skin colour are not the same thing.

1994

Herrnstein and Murray's *The Bell Curve* argues that there are heritable average IQ differences between racial groups

2007

James Watson retires after controversial claims about African intelligence

The evidence for these ideas, however, is weak. While ethnic variation in IQ scores is real, these can readily be explained by socioeconomic factors. Recent data show African-Americans are closing the IQ gap as their living standards rise. The Ashkenazi intelligence hypothesis, while elegant, lacks supporting data. The fact that intelligence is influenced by genes in no way implies that these genes vary in abundance with ethnicity. Any claims that they do must be filed under conjecture, not fact.

Is race scientifically meaningful? Many scientists would go further, and question whether racial differences in any trait are worth investigating. In 1972, American biologist Richard Lewontin argued that the concept of race has neither social value nor scientific significance, and the mapping of the human genome has since convinced many people that he was right. Every human being shares between 99.7 and 99.9 per cent of his or her DNA with every other person on the planet, depending on the measure used. The tiny variations that make individuals unique also differ more within ethnic groups than between them. An African's genes will often look more like those of Caucasians or Chinese than those of another African. The Lewontin doctrine says science should have no business with these misleading categories, which serve only to perpetuate stereotypes.

Evolution of skin colour

Homo sapiens evolved in Africa, and the first humans were probably dark-skinned. Why, then, are so many races pale? A possible explanation is that it is an adaptation to living at higher latitudes. High levels of the pigment melanin protect the skin against damage from the Sun's ultraviolet light, which can cause cancer. Melanin, however, also inhibits production of vitamin D when the Sun's rays are not strong.

As people migrated north of the equator, natural selection may thus have favoured those with lighter skins, because skin cancer became a less serious danger than a lack of vitamin D. This is supported by the way dark-skinned people living in the northern parts of Europe and America have a higher incidence of rickets, a bone disorder caused by vitamin D deficiency. Recent research has even identified a gene, called slc245a5, which may contribute to skin colour.

Genetic discoveries have certainly refuted pseudoscientific racial ideologies – though racial discrimination would still be wrong had science found greater differences between populations. The suggestion that race is biologically meaningless, however, goes too far. Although skin colour is a poor marker for ancestry, fluctuating widely within groups we call 'black', 'white' or 'Asian', it is possible to predict more narrowly defined ancestries from individuals' genomes. These correspond quite well with self-defined ethnicity, and this understanding can have scientific and medical importance.

Certain racial groups have a higher incidence of particular diseases. Sickle-cell anaemia is more common among people of African and Mediterranean backgrounds, multiple sclerosis is more prevalent among Caucasians, and Tay–Sachs disease predominantly affects Ashkenazi Jews (see Chapter 39). This knowledge can be valuable in diagnosis – though doctors must take care not to rule out diseases because a patient belongs to the 'wrong' ethnic group.

Race can also be useful in predicting responses to particular medicines. The antipsychotic drug clozapine is more likely to trigger serious side-effects among people of Afro-Caribbean origin, and the heart drug BiDil has been licensed only for America's black community. In neither case does skin colour matter – but it is often inherited together with other, as yet unknown genes that affect the way these compounds are metabolized. It should eventually be possible to test directly for these genes and prescribe accordingly, but for now, race is a helpful proxy.

Haplotypes – the blocks in which DNA is inherited – also vary with ethnicity, and understanding how this occurs is critical to identifying genes that cause disease. The HapMap, introduced in Chapter 19, includes four ethnic groups – European, Yoruba Nigerian, Han Chinese and Japanese – so that genetic research can cover different populations.

Genetic diversity is indeed greater within races and between individuals than it is between ethnic groups, and there is certainly nothing in the human genome to justify racial discrimination. It is always wrong, in any case, to categorize individuals according to the average characteristics of groups to which they belong. It is misleading, though, to infer from this that genetic diversity between populations does not matter at all.

the condensed idea
Race isn't meaningless

25 Genetic history

Chris Stringer, Natural History Museum, London: 'All men and women are Africans under their skin. It's as simple as that.'

When Charles Darwin wrote *The Descent of Man* in 1871, scientific racism was in its heyday. People of white European origin had come to dominate the globe, and this was widely thought to reflect their superior biological status. Many intellectuals considered humanity to be not one species but several, and Darwin's ideas led some to conclude that dark-skinned people had been left behind by evolution. The idea that we are all Africans would have struck fashionable opinion as preposterous. That, however, is precisely what Darwin suggested in his second great book. As our closest animal cousins, the chimpanzees and gorillas, are native to Africa, he reasoned that the same is probably true of our own species, *Homo sapiens*.

It was a prescient insight. Within 50 years, fossil discoveries had started to point towards an African origin for humanity, and this has now been overwhelmingly confirmed by genetic research. DNA has not only shown that people are all closely related, and far more alike than different. It has allowed us to trace the history of our own species and others, and even to identify some of the biological idiosyncrasies that make us human.

Out of Africa Many of the most significant fossils belonging to ancient human relatives, and all of those that are more than about two million years old, have been discovered in Africa. Fossils such as Lucy, the celebrated *Australopithecus afarensis* specimen unearthed by Donald Johanson in Ethiopia in 1974, have left little doubt that the lineages of humans and chimps separated south of the Sahara.

The more recent evolutionary history of *Homo sapiens*, however, has been less clear. Other human species, such as *Homo erectus* and Neanderthal

c.7 million years ago
Divergence of human and chimp family trees

c.3.2 million years ago
Lucy, the best-known example of *Australopithecus afarensis*, lived

man, had spread beyond Africa well before anatomically modern humans appeared around 160,000 years ago, and two hypotheses have competed to explain the origin of our species.

The 'out of Africa' theory contends that we evolved just once, in Africa, and then migrated to displace our relatives on other continents. The multi-regional view claims instead that pre-existing populations of proto-humans evolved separately, or at least interbred with travelling *Homo sapiens* bands, to give rise to modern races.

The fossils have always favoured 'out of Africa', but genetics has supplied clinching evidence. Two types of human DNA have proved particularly instructive. While most chromosomes are constantly shuffled by recombination, this process does not apply to the genes held by the male

Are we Neanderthals?

A great controversy in human evolution has been the Neanderthals' place on our family tree – did these ancient inhabitants of Europe die out when *Homo sapiens* reached the continent, or were they at least partially assimilated through interbreeding?

Sufficient genetic material has now been recovered from Neanderthal fossils to allow the species' genome to be sequenced, and the results have settled the debate. Modern humans do not appear to have any Neanderthal DNA. If any of our ancestors mated with Neanderthals, their offspring did not survive to contribute to the human genome as it exists today.

Another surprising insight from the Neanderthal genome is that the species had the same version of the FOXP2 gene that modern humans have today. This could mean that they were capable of language, and that they were not the grunting brutes that popular culture often portrays them to have been.

c.2 million years ago
Homo erectus first migrated out of Africa

c.160,000 years ago
Emergence of anatomically modern *Homo sapiens*

c.70,000 years ago
Migration of *Homo sapiens* out of Africa

Y chromosome, and by the mitochondria, which are transmitted from mothers to their offspring. Both are inherited intact, and vary only because of spontaneous mutations.

As such mutations occur at a fixed rate, the DNA of living people can be used to reconstruct their ancestry. And the evolution of Y chromosomal and mitochondrial DNA has progressed exactly as the out of Africa theory predicts. It even charts the routes by which *Homo sapiens* populated the globe.

Further evidence comes from genetic diversity. The out of Africa theory suggests that several thousand people were living in the continent about 70,000 years ago, when a small group crossed the Red Sea, The descendents of these migrants then went on to populate the rest of the world. Non-Africans, who all originate from this small founding group, should therefore be less genetically diverse than Africans, who sprang from a population that was larger and more varied to begin with.

Once again, this is just the pattern that DNA reveals. Human genetic diversity is much greater within Africans than between Africans and any other ethnic group – or even between other ethnic groups that appear closely related. On a genetic level, a Finn will resemble some Africans more than he resembles a Swede. The variability of human DNA declines with increasing distance from the motherland – Australian Aborigines and Native Americans are the least diverse populations of all. Genetic reconstruction techniques are so good that we even know roughly how many people – about 150 – left Africa in that critical first wave.

Whither human evolution?

If humans are themselves the results of past evolutionary forks, could *Homo sapiens* itself evolve into different species? In 2007, research led by American anthropologist Henry Harpending suggested that the answer might be yes. Genetic differences between population groups, he found, had widened over the past 10,000 years. Left to continue, this could potentially result in two or more new species.

Harpending's study, however, looked at the pre-industrial world, when ethnic groups were usually separated by distances too great to travel. Now that air transport and globalization have broken down many geographical boundaries, most evolutionary biologists think a fresh human speciation event is highly unlikely.

What makes us human? Similar methods can be used to chart the evolutionary history of any species and to establish genetic relationships between them. Molecular evidence, for example, shows that the closest living relatives of whales and dolphins are hippos. DNA proves the fact of evolution as surely as the fossil record. Genetic comparisons are also capable of pinpointing some of the evolutionary events that may have been important in the development of particular species. In our own case, they have highlighted at least a few of the genes that appear to make us human.

FOXP2, which we met in Chapter 13, is a prime example. Across mammals and birds, this gene is highly conserved – the sequence is almost exactly the same from species to species, which usually means it does something important. In mice and chimps, which last shared a common ancestor at least 75 million years ago, the FOXP2 protein differs by just a single amino acid.

Humans and chimps diverged much more recently – about 7 million years ago – yet our FOXP2 protein differs by two amino acids from the chimp version. Twice as many mutations as separate chimps and mice have accumulated, in less than a tenth of the evolutionary time. This pattern suggests that natural selection is at work, preserving useful changes. In this case, it may be language: people with FOXP2 defects have major speech impairments. These mutations could be part of the explanation for a capacity that is unique to humans.

> **‘You get less and less variation the further you go from Africa.’**
>
> **Marcus Feldman, Stanford University**

Another stretch of DNA, called HAR1, shows signs of even stronger selection. It is 118 base pairs long, and in the 310 million years since chimps and chickens shared an ancestor, just two of these have changed. Human HAR1, however, is spelt differently from the chimp version in no fewer than 18 places. The rapid pace of its evolution has led scientists to speculate that it might be involved in brain size and intelligence – the most striking difference between humans and other animals. It may be one of the genes that makes us human.

the condensed idea
DNA is a historical record

26 Genetic genealogy

Spencer Wells, director of the Genographic project: 'The greatest history book ever written is the one hidden in our DNA.'

Though membership of the Jewish community is passed on through the female line, orthodox and conservative traditions give special status to a group of men known as the kohanim. In the Book of Exodus, God conferred the title of kohen on Aaron, the high priest and brother of Moses, as an 'everlasting office' that would be passed on to all his heirs in the male line. Every modern kohen thus claims to be a direct male descendent of Aaron, a member of a patrilineal priestly caste who is given particular responsibilities during acts of worship.

In the mid-1990s, a Canadian doctor named Karl Skorecki, who is a kohen, realized that if all kohanim indeed have a common ancestor, albeit more than 3,000 years ago, they should share genetic similarities. The Y chromosome – the package of DNA that determines male sex – is passed from father to son. Could Aaron's Y chromosome, Skorecki wondered, be preserved in today's kohanim?

To find out, he approached Michael Hammer, a geneticist at the University of Arizona who researches the Y chromosome. The pair recruited 188 Jewish men, who gave a sample of their DNA and details of their Jewish heritage. The results were remarkable. Of the 106 who identified themselves as kohanim, 97 shared an array of six genetic markers on the Y chromosome. Most really had a common male ancestor in the distant past. A genealogical tradition had been confirmed by molecular genetics.

1991
Mitochondrial DNA identifies bodies of Tsar Nicholas II's family

1997
Identification of kohen Y chromosome

Gene trees Genetic genealogy has since become big business. For a fee of a few hundred pounds, dozens of companies will now test your DNA to assess your ancestry. The most useful tool, for men at least, remains the Y chromosome. As we saw in the last chapter, this is not scrambled by recombination in every new generation. Like Anglo-Saxon surnames, it thus tumbles down the male line more or less intact. By looking at its mutation rate, scientists can place men into groups who share a long-dead male ancestor.

There are 18 broad Y-DNA clans or 'haplogroups', the origins of which have been traced to particular geographical regions. Haplogroups A and B are exclusively African, haplogroup H originated in the Indian subcontinent, and haplogroup K is peculiar to the Aborigines of Australia and New Guinea. Many can be broken down into narrower subgroups. R1b is most common among European men, while the kohanim belong to J1 and J2. Aaron, it seems, may have lived sufficiently long ago for his male line to have split in two.

Women, of course, do not have Y chromosomes, but female gene trees can be drawn up using mitochondrial DNA (mtDNA), which both men and

Crusaders and Muslims

Historical events have often left a genetic legacy that is visible in the DNA of people alive today. A recent study of modern Lebanese has revealed that a disproportionate number of Christian men have a Y chromosome that is clearly of Western European origin. This was probably carried to the region by the Crusaders between the 11th and 13th centuries, and has since been passed on by the descendents of those who settled in the region.

The research also found that a Y chromosomal type with roots in the Arabian Peninsula is more common among Lebanese Muslims, which may reflect earlier migrations during the Islamic expansion of the 7th and 8th centuries.

2001
Brian Sykes publishes *The Seven Daughters of Eve*

2005
Launch of Genographic project

women inherit from their mothers, and which also escapes recombination. People can thus be separated into almost 40 matrilineal haplogroups, which have again been linked to different parts of the globe.

Anybody can therefore learn something about his or her ancient ancestry with DNA tests, which genetic genealogy companies are now marketing in imaginative ways. Oxford Ancestors, set up by the geneticist Bryan Sykes, uses mtDNA to assign Europeans to clans founded by the 'Seven Daughters of Eve' – putative matriarchs with names such as Ursula (for haplogroup U) or Helena (for haplogroup H). The company also specializes in linking men's Y-DNA to the great pillagers of history, the Vikings and the Mongols. Sykes has even claimed that one of his clients, an American accountant named Tom Robinson, has the Y chromosome of Genghis Khan.

Such DNA tests are also useful for drawing more recent family trees. Genetic techniques can confirm relationships, which can be of great value to historians both amateur and professional. When the bodies of Tsar Nicholas II and his family were exhumed in 1991, genetic testing was used to confirm their identities. The Tsarina Alexandra was matched to an mtDNA sample provided by her great-nephew, the Duke of Edinburgh.

The Genographic Project

The world's biggest exercise in genetic genealogy is the Genographic Project, a $40 million collaboration between National Geographic and IBM launched in 2005, which is seeking to collect at least 100,000 DNA samples from indigenous people around the world. The aim is to reconstruct the history of human migrations, and to investigate the genetic relationships between different ethnic groups. It has also sold more than 250,000 personal genetic testing kits at $100 each to allow individuals to trace their ancestry.

Like a similar initiative, the Human Genome Diversity project, it has drawn criticism from some geneticists and from organizations representing indigenous peoples, who argue that it could promote racism by identifying genetic signatures for particular ethnic groups. There is also some concern about obtaining informed consent from societies with little understanding of genetics.

Amateur genealogists who find fellow enthusiasts with the same surname can learn whether they are related through Y-DNA tests, helping both parties to extend their family trees. In 2005, a 15-year-old boy who had been conceived by sperm donation even used an online genetic database to find his biological father. His Y chromosome matched that of two men with the same surname, and as the boy's mother already knew the donor's date and place of birth, this led him to the right man.

> **'Of the literally thousands of genetic ancestors you had 12 generations ago – say about the year 1700 – mitochondrial DNA is connecting you with only one.'**
>
> **Jonathan Marks, anthropologist**

Buyer beware The value of many commercial tests has been questioned by professional geneticists. In the US, these are particularly popular among African-Americans seeking to trace their roots: television personality Oprah Winfrey claimed recently that her DNA shows she is a Zulu. This is almost certainly misleading. Even if her mtDNA belongs to the same haplogroup as most Zulus, that says very little about her ancestry. Go back just 20 generations, and we all have more than a million direct ancestors. Oprah's mtDNA test tells her who just one of these people might have been. The same applies to the Y chromosome. Tom Robinson could be descended from any Asian male who had lots of sons and grandsons – there is no evidence that it was Genghis Khan.

DNA tests can also throw up uncomfortable results. Many male African-American clients have been surprised to discover that their Y chromosomes belong to typically European haplogroups – a legacy of plantation-owners' sexual exploitation of their female slaves. Such tests, indeed, have confirmed that Thomas Jefferson, the third US president, fathered children by his slave Sally Hemmings. Comparisons of genetic material from the confirmed male descendents of both show the same Y-DNA.

Genetics, too, can disprove biological relationships as well as confirm them. Many geneticists tell anecdotes about having to exclude people from family studies as their DNA shows unequivocally that they are not related to the men they believe to be their fathers. Genetic genealogy can be fun, and in the right context, historically illuminating. But it won't show you're a Viking or a Zulu, and it may hold nasty surprises.

the condensed idea
Genes can trace ancestry

27 Sex genes

Steve Jones: 'In its brief moment of glory, [SRY] sends billions of babies on a masculine journey.'

Eve, according to the Bible, was fashioned from Adam's rib. Yet if genetics has surprised racists by revealing that the cradle of humanity is Africa, it has also surprised male chauvinists. DNA has revealed that the Book of Genesis got the story the wrong way around. By default, human beings are genetically programmed to be female.

In the movie *My Fair Lady*, Professor Henry Higgins famously asked: 'Why can't a woman be more like a man?' But from a genetic perspective, the question is more interesting and revealing when posed in reverse.

Why can't a man be more like a woman? The discovery of the underlying genetic reason for the differences between the sexes was made independently, and appropriately, by a woman and a man. In 1905, Nettie Stevens and Edmund Beecher Wilson noticed that male and female cells varied in chromosomal structure. While females had two copies of a large chromosome, the X, males had just one, along with another, much smaller chromosome, the Y. They had identified the system by which sex is determined in many animals, including humans: women have the chromosomal type XX, while men are XY.

When meiosis separates pairs of chromosomes to create gametes with a single set, eggs always carry an X, while sperm can carry an X or a Y. X-bearing sperm will produce girls when they fertilize eggs; Y-bearing sperm boys. For the first six weeks of gestation, both male and female embryos develop in identical fashion. They would continue to do so, to produce babies that look female, without the action of a single Y chromosome gene. A woman's extra X chromosome sends no extra signal that makes her into a woman. That is the way people will be, without the intervention of a gene called SRY.

timeline

1905
Nettie Stevens (1861–1912) and Edmund Beecher Wilson (1856–1939) identify sex chromosomes

1910
T.H. Morgan discovers sex-linked inheritance

The masculinity switch SRY, which was discovered in 1990 by Robin Lovell-Badge and Peter Goodfellow, stands for sex-determining region Y, and it is the biological key to sex. Those people who have a working copy will grow a penis, testicles and beard, while those who lack one will develop a vagina, womb and breasts. It is perhaps the most influential single gene in the human body.

If it does not kick into action seven weeks into pregnancy, or if its instructions cannot be heard, the body will continue to develop along its default female path. Should an XY embryo's SRY gene be mutated and non-functional, or should other genetic faults make cells insensitive to the male hormones the gene directs the gonads to produce, that embryo will grow into a girl (though she will be infertile). On rare occasions, SRY can find its way into an X chromosome, by means of a kind of mutation called a translocation. It will not surprise you to learn that when that happens, XX individuals become male.

Sex selection

The chromosomal differences between men and women mean it is now possible to choose the sex of one's children. The most effective method is to create embryos through in-vitro fertilization, and then remove a single cell to check whether it has two X chromosomes or an X and a Y. Only embryos of the desired sex would then be transferred to the womb.

Another method, known as MicroSort, relies on the different sizes of the X and Y chromosomes. Sperm are treated with a fluorescent dye that stains DNA, and these are then passed under a laser. As the X is so much larger than the Y, X-bearing sperm will glow more intensely, and can be separated out. The technique is claimed to increase the chances of having a child of the desired sex to between 70 and 80 per cent. It is permitted in the US, but not in the UK.

1990
Discovery of SRY gene by Robin Lovell-Badge and Peter Goodfellow

2003
Simon Baron-Cohen publishes systemizing and empathizing brain hypothesis

SRY works as a masculinity switch. After five weeks of pregnancy, all embryos start to develop unisex gonads, which have the potential to become testes or ovaries. Then, two weeks later, the SRY switch is flipped, or not. Once turned on, it tells these gonads to become testes. If it is absent, switched off, or silent, these gonads will later begin developing into ovaries.

After eight weeks, the newly sculpted testes start to make male hormones, and these androgens masculinize the body. Clusters of cells that would otherwise become the clitoris and labia form the penis and scrotum, and the sex organs are plumbed together using ducts that atrophy in females. SRY makes a man.

Sex differences SRY is not only the root cause of the physiological differences between the sexes. It also plays a part in those behaviours that are more common among people with a Y chromosome, such as risk-taking and aggression. None of these is directly programmed by SRY, though some of the human Y chromosome's other 85 or so genes might be associated with traits that are more commonly observed among men. But

Masculinity and health

No common gene is more damaging to health than SRY. Women outlive men in every society, and one of the reasons is the hormonal profile that this male gene creates. High testosterone levels make men more likely to take risks that endanger their survival, whether through careless driving, aggressive behaviour, or smoking and drug use. The female hormone oestrogen also protects against cardiovascular disease, which is the biggest killer of both men and women. Alzheimer's is the only major disease that commonly affects both sexes for which women have a higher risk.

Being male is also a major risk factor for autism, which is four times more common among boys than girls. Simon Baron-Cohen has suggested that this could be related to excess pre-natal androgen exposure, which creates an 'extreme male brain' that often excels at systemizing, but has little empathizing ability.

they nonetheless fall under the influence of this powerful gene. The cascade of androgens that SRY sets off masculinizes minds as well as bodies.

This indirect genetic effect is probably at least as responsible for typically male personality traits as are culture and learning. Men's higher testosterone levels certainly make them more prone to violence and recklessness, and they may also affect personality.

Simon Baron-Cohen, of the University of Cambridge, has suggested that one example is the way that women tend to be better than men at empathizing – at identifying others' thoughts and emotions, and then acting appropriately. Men, on average, are better at systemizing – at building and understanding systems, like car engines, mathematical problems and football's offside laws.

Baron-Cohen's work hints that this may be related to androgen exposure in the womb. In one study, his team examined pre-natal testosterone levels in 235 expectant mothers who had an amniocentesis test to check for foetal abnormalities, and then followed up the children when they were born. Those exposed to more testosterone tended to look less at other people's faces, and to acquire stronger numerical and pattern-recognition skills.

This research is easily misunderstood. It in no way suggests that it is any better to be a 'systemizer' or an 'empathizer', or that either trait is associated with greater intelligence. Neither are all men created one way and all women another. It is only on average that more men will have the first brain type and more women the second, just as men are on average taller than women while some women are taller than some men.

These averages, however, are part of a growing understanding that men and women are not biologically identical in their thought processes and behaviour, any more than in their reproductive systems. And the root cause of most of these differences is a single Y-chromosome gene.

the condensed idea
Men are genetically modified women

28 The end of men?

Bryan Sykes: The human Y-chromosome is crumbling before our very eyes.'

The Y chromosome is the runt of the human genome. While its counterpart, the X, has more than 1,000 genes, including many that are critical to metabolism in both sexes, the Y has fewer than 100. It was once identical to the X, but since it began to change around 300 million years ago, it has progressively shrunk to the point at which it holds less genetic information than any other chromosome. Chromosomes 21 and 22 are smaller in size, but each carries many more genes.

It is also a loner. While the X chromosome pairs up in female bodies, the Y always leads a solitary life among men. This isolated existence can be medically damaging. As women have two X chromosomes, one of which is inactivated in each cell, they have a ready-made spare should one of its genes become mutated. If it is an essential one, like the dystrophin gene involved in muscle development, the other X can compensate and the woman remains healthy.

Men are not so lucky. Their second sex chromosome is the Y, which has been shorn of most genes. It can offer no such back-up. If the solitary X bears a mutated dystrophin gene, the result is Duchenne muscular dystrophy – a crippling wasting condition that confines boys to wheelchairs by childhood, and kills them by their 20s by paralysing the muscles they need to breathe. Many other life-threatening conditions, such as haemophilia and severe combined immune deficiency, are similarly X-linked, and mostly affect boys. Women can be carriers, but hardly ever inherit the two mutated X chromosomes they would need to fall ill.

timeline

***c.*300 million years ago**
Split between human X and Y chromosomes

1905
Discovery of sex chromosomes

The descent of men X-linked diseases are not the only way in which men's split chromosomes leave them at a disadvantage. The absence of a counterpart in the human genome also prevents the Y from taking part in recombination, the process that allows the other chromosomes to protect themselves against mutation and decay. As we saw in Chapter 6, when cells divide by meiosis, their paired chromosomes exchange blocks of DNA. This allows them to escape Muller's ratchet – the process by which deleterious mutations would otherwise accumulate in each generation, causing irreversible long-term decay.

While the X chromosome can recombine with the other X with which it is paired in women, most of it cannot recombine with the Y. Though the X and Y were once the same chromosome, evolution has gradually removed their ability to swap DNA with one another. Some genes on the Y would have been harmful if inherited by females, and vice-versa: were SRY to

Inherited infertility

More than half of in-vitro fertilization procedures now involve a new technique called intra-cytoplasmic sperm injection (ICSI), in which a sperm is injected directly into an egg to fertilize it. It has revolutionized the treatment of male infertility, as it allows men whose sperm are too weak to swim to an egg and penetrate it to become fathers. Even men who ejaculate no sperm at all can be helped: non-swimming sperm with no tails can be surgically removed from the testis and injected to create an embryo.

ICSI, however, may have a drawback. If these men are infertile because of mutations or deletions on the Y chromosome, injection of their non-swimming sperm will probably pass their fertility problems on to their male offspring. Few people think this makes the technique ethically dubious – these boys, if infertile, will be able to use ICSI themselves. But it shows how medical science has allowed something that is impossible in nature – the inheritance of infertility.

1990
Discovery of SRY gene

2003
Y chromosome sequence reveals gene conversion

2003
Bryan Sykes publishes *Adam's Curse*

cross over to the X, for instance, it would make females male. Certain genes on the X chromosome are also critical to healthy development in either sex. Recombination would have left some males without these essential parts of the genome.

Denied a partner with which to recombine, the Y has decayed in more or less the fashion predicted by Muller's ratchet. Every mutation that was not fatal to the man who carried it, or to his ability to reproduce, has been preserved in his male descendents. While this has been a boon to genealogists, allowing the ancestry tracing explored in Chapters 25 and 26, it has been bad news for the Y's complement of genes. More and more have been progressively knocked out by mutation, leaving behind the withered chromosomal shell that survives today.

This story of genetic corrosion can be expected to continue, so that more genes will be lost from the male chromosome. This led Jenny Graves, an Australian geneticist, to propose that the Y chromosome is slowly but surely on its way out. The idea has been popularized by British geneticist Brian Sykes in his book *Adam's Curse* (2003), which suggests that at the present rate of decline, men have only another 125,000 years left. He

The mole vole

Another way in which the male sex may yet survive the degenerative forces ranged against the Y chromosome is illustrated by a small rodent native to the Caucasus mountains – the mole vole. The male of this species has lost its Y chromosome altogether, but it remains male all the same.

Though both the Y and its critical payload, the SRY gene, are gone in the mole vole, it has evolved a cobbled-together alternative. The job of sex determination has been switched to another chromosome, which seems to activate a 'gene relay' that is normally started by SRY. Bryan Sykes even suggests that this might be done by genetic engineering, to create an artificial 'Adonis chromosome' that transmits maleness without the weaknesses of the Y. Such a chromosome, however, would be expected to decay in due course, just like the Y has done.

prophesies that the end of men is nigh – and with it, perhaps, the end of humanity.

Escaping Adam's curse Most scientists do not share Sykes's pessimism. First, there is natural selection to think of. Some of the Y chromosome's genes are important in sperm production. Spontaneous mutations that inactivate these should therefore eliminate themselves from the gene pool, by reducing male fertility (see box for a modern exception). The same calculus applies to the Y's signature gene, SRY. If it is mutated and non-functional, an embryo will grow into an infertile woman with no uterus or ovaries. Its centrality to masculinity and sexual reproduction makes it immune to Muller's ratchet: harmful variants might arise, but they cannot spread through the gene pool because carriers cannot breed.

'If 1 per cent of men are infertile because of problems in the Y-chromosome, you still have 99 per cent of normal fertile men. Nature tends to eliminate the Y-chromosomes that reduce fertility.'

Robin Lovell-Badge

This is not the only objection to the 'end of men' thesis. The Y chromosome, it turns out, has also evolved a unique way of repairing mutations. When the sequence of the Y chromosome was completed in 2003, it showed much less of the genetic decay than was expected from 300 million years of Muller's ratchet. But it also emerged that large parts of its code are written in palindromes. Like 'Madam, in Eden, I'm Adam', these read the same way when reversed. Such palindromic phrases reach up to 3 million base pairs in length.

They exist for a reason: they protect the Y's genome, and allow it to repair errors. When Y chromosome genes are replicated, a process called gene conversion takes place. New copies are effectively checked against the mirror image contained in a palindrome, to eliminate mistakes. Instead of recombining with a partner chromosome, the Y recombines with itself. As Steve Jones, of University College London, has put it: 'If so, its salvation lies in that very masculine habit of sex with oneself.'

Men may owe their nature to a reclusive and degenerate chromosome, but rumours of its demise – and theirs – have been greatly exaggerated.

the condensed idea
Men are genetically degenerate

29 The battle of the sexes

Matt Ridley: 'To turn anthropomorphic, the father's genes do not trust the mother's genes to build a sufficiently invasive placenta, so they do the job themselves.'

In 1532, on a visit to the town of Dessau, Martin Luther was presented with a child that behaved so strangely that he doubted its humanity. 'It did nothing but eat; in fact, it ate enough for any four peasants or threshers,' the father of the Reformation noted. 'It ate, shat, and pissed, and whenever someone touched it, it cried.' Luther thought the diagnosis simple: the child was possessed by the Devil, and should be thrown into the river Molda to drown. 'Such a changeling child is only a piece of flesh, a *massa carnis*, because it has no soul,' he said.

Today, we would come to a different conclusion. From the symptoms described by Luther's chronicler, Johannes Mathesius, paediatricians would immediately suspect Prader–Willi syndrome. Far from having no soul, what it lacked was probably a genetic region called 15q11, which when missing causes the grotesque overeating, floppy muscles and learning difficulties that were described in 1956 by Andrea Prader and Heinrich Willi.

Luther's changeling must have inherited its 15q11 mutation from its father. We can say that with confidence because had the defect occurred in its maternal copy of chromosome 15, it would have developed an entirely different disease. In 1965, an English doctor named Harry Angelman reported three rare cases of what he described as 'puppet children'. They were small and thin, they moved stiffly and jerkily, they were severely mentally retarded, and they had an unusually happy demeanour.

timeline

1956
Identification of Prader–Willi syndrome

1965
Identification of Angelman syndrome

Incredibly, their condition is caused by precisely the same stretch of DNA as is Prader–Willi.

Imprinted genes The genetic disease that a child with a 15q11 defect will get depends on which parent supplied the mutated chromosome. If it came from the mother, it will be Angelman syndrome, and if it came from the father, it will be Prader–Willi. The gene that underlies these disorders is imprinted – that is, it carries a biological marker that tells cells to express only the maternal or paternal copy. Imprinted genes can 'remember' their parental history, through a process known as methylation that leaves some switched on and others switched off (see box).

Dozens of imprinted genes are now known, and a large proportion of them are involved with the development of embryos. Imprinting seems to require that a viable embryo has genetic input from both a man and a woman. This is superficially obvious – conception occurs when a sperm fertilizes an egg, so of course both sexes are involved. After fertilization, however, the pronuclei of the two gametes do not fuse immediately, and the sperm pronucleus can be switched for one from an egg, or vice versa. Thus can scientists create embryos with two genetic fathers or mothers.

Methylation

Imprinting works because of a process called DNA methylation, by which the function of genes is altered by chemical modifications. It involves the addition of a chemical tag, known as a methyl group, to the DNA base cytosine. This can turn down a gene's activity, or switch it off entirely. It is critical to ensuring that genes are expressed only at the right times in an organism's life cycle, and in the right kinds of tissues.

Most of these methyl tags are wiped away during the early stages of embryonic development. The main exceptions are the imprinted genes, which retain these marks to flag up their maternal or paternal origin.

1980s
Mouse experiments show maternal and paternal genes are necessary for embryonic development

1990s
David Haig proposes influence of imprinted genes on placenta

In principle, such embryos should develop normally – they have a full complement of chromosomes and the cellular architecture they need to grow. But they don't. Experiments with mouse embryos have shown not only that they fail and die, but that they fail in different ways according to the source of their genes.

When all the genetic material is female, the inner cell mass that will eventually become the foetus starts to form normally, but dies because it lacks a viable placenta. When the embryo has two genetic fathers, the placenta forms normally, but the inner cell mass is a shapeless mess – a *massa carnis*, as Luther might have called it. Both sexes are necessary: imprinted paternal genes are essential for a healthy placenta, while imprinted maternal genes are needed to organize the embryo.

Artificial gametes

An exciting implication of stem cell research (a field described more fully in Chapter 35) is the prospect of growing artificial eggs or sperm, to allow men and women who make none to have their own genetic children. This possibility, however, has also led to speculation that sperm could be made from female cells, or eggs from male ones, to allow gay couples to conceive. It has even been suggested that the same person could produce both sets of gametes, in the ultimate form of self-love.

Imprinting, however, suggests that making working 'male eggs' or 'female sperm' will be very difficult. It would be necessary to ensure they carry all the right tags to denote maternal or paternal genes, the full range of which remains unknown. Sperm also require a Y chromosome, which female cells lack.

Imprinting issues are also thought to explain the developmental problems suffered by many animal clones. Another likely effect is the inability of mammals, unlike bees, lizards and sharks, to reproduce by parthenogenesis – a process by which eggs develop spontaneously into embryos without fertilization.

A hostile takeover The evolutionary reasons for this can be understood by thinking of the placenta as a foetal organ – an idea first proposed by Australian biologist David Haig in the early 1990s. It is a tool by which the foetus becomes a kind of parasite, effectively stealing from its mother the nutrients, oxygen and the other resources it needs to grow.

The interests of the foetus and its mother are therefore slightly different: while the foetus benefits by squeezing as much out of its mother as it can without killing her, the mother tries to hold some resources back for the sake of her health. This leads to a uterine tug of love, which may be responsible for pregnancy complications such as pre-eclampsia and gestational diabetes.

‘The phenomenon is called imprinting because the basic idea is that there is some imprint that is put on the DNA in the mother's ovary or in the father's testes which marks that DNA as being maternal or paternal, and influences its pattern of expression — what the gene does in the next generation in both male and female offspring.’

David Haig

This means that though half a foetus's genes come from its mother and half from its father, each set has different interests. Maternal genes will benefit from being less demanding, so the woman they came from has a good chance of breeding again. To paternal genes, however, the mother's future reproductive prospects do not much matter. They benefit from diverting as much of her investment as possible to the current foetus: she might conceive her next one with someone else. Imprinted genes from the father therefore create an aggressive placenta, which stages something of a hostile takeover of the mother's womb.

Haig's hypothesis has been supported by subsequent discoveries about imprinted genes and embryonic development. The gene for an insulin-like growth factor called IGF2, for example, is active in building the placenta, but switched off in adults. It is also paternally imprinted. A gene that appears to counteract its effects, called H19, is imprinted too, but from the maternal side.

Imprinted genes are not widely found in egg-laying animals: as their embryos are not fed through a placenta, and so cannot influence the resources they get, there is no need for maternal and paternal genes to fight it out. The placenta is not just a highly efficient means of nourishing offspring. It is also the *casus belli* for a fierce battle of the sexes.

the condensed idea
Natural selection forms new species

30 Homosexuality

Dean Hamer: 'I suspect the sexual software is a mixture of both genes and environment, in much the same way the software of a computer is a mixture of what's installed at the factory and what's added by the user.'

In the 1960s, a piece of graffiti is said to have appeared in a London public lavatory. Underneath the slogan 'My mother made me a homosexual,' another hand had written: 'If I give her the wool, will she make me one too?'

The story may well be apocryphal, but it illustrates a widely held belief about homosexuality: that it is conditioned by people's environment, experience and upbringing. It is a view that is popular among religious conservatives, who regard the practice as a sinful personal choice. But it also appeals to some homosexuals, who think anybody is potentially gay given the right social context, or fear that the discovery of biological causes might be used to develop a 'cure'.

Other gay men and lesbians, however, feel certain that they were 'born that way'. While few scientists would deny that nurture is involved in sexual orientation, there is ample evidence that biology, perhaps including genetics, is implicated too. Homosexuality occurs in every known human culture – always a strong indicator of a natural phenomenon. Identical twins are also more likely to share their sexual orientation than are fraternal sets, another sign that genes probably play a role.

A gay gene? In the mid-1990s, a T-shirt popular on the gay scene bore the legend: 'Xq28. Thanks for the genes, Mom!' It celebrated the work of American geneticist Dean Hamer, who in 1993 claimed the discovery of the first gene linked to homosexuality. Hamer had noticed that gay men often have gay male relatives, but chiefly on the maternal

timeline

1932
J.B.S. Haldane suggests kin selection effects could explain survival of homosexuality despite its evolutionary cost

1993
Dean Hamer (b.1951) links Xq28 genetic region to male homosexuality

side of their families. He realized that this might implicate the X chromosome, which men always inherit from their mothers, and began to compare the X chromosomes of gay and straight men.

Of 40 pairs of gay brothers in the study, 33 shared a particular set of variants in a region called Xq28. Straight brothers tended to share a different set of variants at the same spot. Xq28, Hamer suggested, might influence male homosexuality. It was not a 'gay gene' *per se* – some men who have the supposedly homosexual alleles are heterosexual, and vice versa. But it was the first plausible candidate for a gene that might predispose men to homosexuality, perhaps in concert with other genetic or environmental cues.

The link to Xq28 has been replicated in some other studies, but not in others. It remains impossible to say with certainty whether it affects male homosexuality or not. As with IGF2R and intelligence, even if it makes a difference, it is almost certainly just one of many genes that have similar effects.

gay animals

The natural origins of homosexuality are vividly illustrated by the antics of the animal kingdom. Bruce Bagemihl, of the University of British Columbia, has shown that same-sex sexual activity is known among at least 1,500 species, and it is well documented for more than 450 of these.

Female bonobos often bond by rubbing their genitals together, and Japanese macaque monkeys can also be enthusiastic lesbians. In some giraffe herds, nine out of ten sexual acts occur between males, up to 8 per cent of rams prefer to mount other males, and penguins, swans and dolphins are known to form same-sex pairs. None of this proves that genes are involved, but the ubiquity of homosexuality is a little suggestive. It is certainly not unique to humans.

1997
Ray Blanchard finds link between male homosexuality and older brothers

2004
Andrea Camperio-Ciani finds gay men's female relatives tend to be more fertile

An evolutionary paradox The possible role of genes in homosexuality raises an interesting evolutionary question. Natural selection is a cruelly efficient eliminator of variants that adversely affect the ability to reproduce, and from a Darwinian perspective, homosexuality is a crime against evolutionary fitness. Even if homosexuals have often married and had children, like Oscar Wilde, mutations that slightly discourage breeding ought never to have spread through the gene pool. How could genes that predispose to homosexuality possibly have survived?

The answer may lie in the effect these genes have on women. If a mutation increases female fertility, so that women who inherit it have more children, it could thrive even if it has the opposite effect on men. Support for this hypothesis has come from the work of Andrea Camperio-Ciani, of the University of Padua, who in 2004 investigated the extended families of 98 homosexual and 100 heterosexual men. He found that the female relatives of homosexuals were significantly more fertile. The gay men's mothers had an average of 2.69 children, compared with 2.32 for the mothers with straight sons, and their maternal aunts had more children, too.

Birth order There is another way in which the evolutionary paradox of homosexuality might be explained. It is conceivable that in large families with many sons, the selfish interests of genes might be better served if younger brothers were to opt out of competition for women, and invest their resources instead in nieces and nephews who share a quarter of

Lesbians

Most research into the biological origins of homosexuality has concentrated on gay men, with little focus on women. Some twin and family studies suggest that lesbianism is at least moderately heritable, and there is evidence implicating higher testosterone levels. No study, however, has yet so much as tentatively suggested the existence of a 'lesbian gene', and elegant evolutionary explanations are also lacking. This may partially reflect greater difficulties in studying lesbians, as more women than men identify as bisexual, or that in male-dominated societies, lesbians have not been free to follow their sexual orientation. But some lesbians have complained that science often seems to pretend they don't exist.

their DNA. This could account for one of the best-attested observations about homosexuality: that it is linked to birth order.

Research led by Ray Blanchard, of the University of Toronto, has established conclusively that men with older brothers are more likely to be gay. The chances increase by about a third with every older brother a man has – though as the background rate of homosexuality is low, most men with many older brothers will still be straight.

This may be caused by the uterine environment. A mother's immune system always reacts against the foetus she carries, as it is genetically foreign. As males have a Y chromosome, which their mothers do not, this response is stronger when the foetus is male, and it increases with every male pregnancy. This may affect hormone profiles, and the sexual development of the brain.

The birth order effect does not apply to boys with older adoptive or step-brothers, which strongly suggests that biology, and not family circumstance, is responsible. So, too, do ring fingers: both gay men and women tend to have long ring fingers, a sign of pre-natal exposure to high levels of testosterone. The link to birth order may be evolved, as suggested above, or it may have persisted despite natural selection. As it primarily affects families with many children, they may not be greatly disadvantaged in evolutionary terms.

Sexual orientation seems to begin with a myriad of intertwining factors – some genetic, some hormonal and gestational, some the product of cultural conditioning. The relative contribution of each remains an open question, and for every individual, the mix may be different. This makes the genetic origins of homosexuality almost impossible to unpick: it is a fair bet that no 'gay genes' will be discovered for which embryos might be screened, or antidotes developed. Gay men and women have little to fear from this field of research. It is establishing that their sexuality is neither a disorder nor a choice, but a part of normal human variation.

the condensed idea
Biology influences sexuality

31 Genetic fingerprinting

Alec Jeffreys: 'There was a level of individual specificity that was light years beyond anything that had been seen before. It was a eureka moment. We could immediately see the potential for forensic investigations and paternity.'

On 2 August 1986, the body of a 15-year-old girl named Dawn Ashworth was discovered in a wood near the English village of Narborough. She had been raped and strangled, in a very similar fashion to Lynda Mann, a 15-year-old from the same village who had been murdered three years previously. Richard Buckland, a 17-year-old who lived locally, was soon arrested. But while he confessed to the second murder, he would not admit to the first.

The police were convinced that the crimes were the work of the same man: the *modus operandi* was identical and semen of the same blood type had been found on both bodies. In search of evidence, officers approached Professor Alec Jeffreys, of Leicester University, a geneticist who had recently developed a method for identifying people from their DNA. He agreed to compare Buckland's DNA against the crime scene samples.

The results were shocking: the girls had been killed by the same man, but it could not have been Buckland. His DNA proved his confession had been false, and the case against him was dropped. The police began taking blood samples from more than 5,000 local men, but no matches were found until a man was overheard boasting that he had taken the test for a friend. The friend, Colin Pitchfork, was arrested, and his DNA was a

timeline

1984
Genetic fingerprinting developed by Alec Jeffreys (b.1950)

1990s
Polymerase chain reaction applied to genetic fingerprinting, allowing smaller biological samples to be tested

perfect match. He confessed, and was jailed for life on 23 January 1988. Genetic fingerprinting had solved a murder for the first time.

The technique The test that convicted Pitchfork relies on repetitive segments of junk DNA known as minisatellites, which are between 10 and 100 letters long. These feature the same core sequence – GGGCAGGAXG, where X can be any of the four bases. Minisatellites occur at more than 1,000 locations in the genome, and in each of these places they are repeated a random number of times.

Jeffreys stumbled on their forensic potential by accident. While studying minisatellites for clues to the evolution of disease genes, he examined DNA samples taken from his lab technician, Vicky Wilson, and her parents. Though the number of minisatellite repeats showed a family resemblance, each profile was unique.

Jeffreys immediately realized the implications: as every person has a highly individual genetic fingerprint, these could be used to match suspects to blood or semen found at crime scenes. Another suggestion came from Jeffreys' wife: the technique could also prove whether would-be immigrants who claimed to be related to British citizens were telling the truth, or to confirm paternity of a child.

Use and abuse DNA fingerprinting has transformed forensic science. It has convicted hundreds of thousands of criminals like Pitchfork – and, just as importantly, has exonerated innocent people like Buckland. Another forensic use has been in identifying bodies. In 1992, it proved that a man buried in Brazil under the name Wolfgang Gerhard was Josef Mengele, the fugitive Auschwitz doctor, and it was used after the 9/11 terrorist attacks to identify victims' body parts.

The actor Eddie Murphy, the film producer Steve Bing and the footballer Dwight Yorke are just three of the thousands of men whose disputed paternity has been confirmed by DNA. The technique even proved that the semen stain on Monica Lewinsky's infamous blue dress contained the 'presidential DNA' of Bill Clinton.

1992
DNA evidence proves identity of corpse of Josef Mengele

1998
Bill Clinton's 'presidential DNA' found on Monica Lewinsky's dress

2003
UK Criminal Justice Act allows DNA to be kept from anybody arrested for an offence, even if never charged or convicted

The technology has advanced considerably since the Pitchfork case. A technique called the polymerase chain reaction (PCR), invented in 1983 by Kary Mullis (see box), was soon incorporated into forensic genetics. As this allows small quantities of DNA to be amplified, it means that as few as 150 cells can make up a readable sample: suspects can now be identified from mere traces of biological material. Microsatellite analysis has also been replaced by the use of smaller repetitive chunks of DNA called short tandem repeats, which are more likely to survive environmental exposure and amplify well with PCR.

Kary Mullis

The inventor of the polymerase chain reaction (PCR) is one of the more colourful Nobel laureates. He has spoken openly of his experiences with LSD, and his autobiography *Dancing Naked in the Mind Field* describes a 1985 incident in which he believes he encountered a talking, glowing raccoon. He has become a controversial figure for other reasons, too: he has supported the maverick claim that HIV does not cause Aids, and championed astrology. The importance of his contribution to molecular biology, however, is unquestioned. As PCR allows DNA to be amplified, it has greatly improved the sensitivity of genetic fingerprinting and genetic testing for disease.

Many countries now routinely store DNA from convicted criminals, and sometimes, as in the UK, even from people who are arrested but never charged. The UK database holds samples from about 4 million people – 6 per cent of the population. As only one in a million people share the same genetic fingerprint, a match to a crime scene sample is often seen by lawyers and juries as conclusive proof. It has even been used by supporters of the death penalty to argue that miscarriages of justice are no longer possible.

Useful as genetic fingerprinting can be, however, its importance is often overstated. First, there is the 'prosecutor's fallacy'. If a profile is held by one person in a million, then in a country of 60 million like the UK, it will be shared by 60 people. Every crime scene sample thus has 60 potential origins that are all equally likely. Unless other evidence points to a suspect, a match means the chance someone is innocent is not one in a million, but 59 in 60.

Another problem is that genetic fingerprints merely place suspects at crime scenes: they provide circumstantial evidence, which may not indicate guilt. It is one thing if a suspect's DNA is found in semen recovered from a rape victim, but quite another if it is found at his local general store after a robbery. If he frequented that shop, his DNA may be present for entirely innocent reasons. Contamination is a further issue: it is possible for an innocent person's DNA to appear at a crime scene because he has opened the same door as the guilty party or shaken hands with him (see box).

Genetic fingerprinting has caught thousands of rapists and murderers, and there is no question that it serves the cause of justice. It is only a tool though, and it is by no means infallible.

Low copy number testing

Contamination is a particular issue for a forensic technique called low copy number testing, which can match genetic fingerprints to DNA from as few as five cells. It is very hard to prove, however, that these come from a guilty person and not an innocent third party.

When you pick up an object, your hands will always deposit a few cells, and pick up others left behind by other people who have handled it. Some of these cells can then be shed onto other surfaces you touch. A frequently touched object, like a door handle, could thus transfer innocent people's DNA onto the hands of a criminal, and thence to a crime scene.

When large biological samples such as semen stains are tested, this is not a problem. The criminal's cells will far outnumber those of third parties, which can be discounted. Tiny samples of just a few cells, however, present a problem: it is hard to be certain they have not been innocently transferred. In 2007, such concerns led to the collapse of the trial of Sean Hoey, who was accused of the 1998 Omagh bombing in Northern Ireland, which killed 29 people.

the condensed idea
DNA reveals identity

32 GM crops

Sir David King, former UK government chief scientist: 'GM is a complex technology, not a homogeneous technology. It has to be considered on a case-by-case basis.'

People have been genetically modifying plants for thousands of years. All the crops that are farmed around the world, from rice to cassava to apples, have genomes that differ markedly from their wild relatives, as a direct result of human intervention. Plants with sweeter fruit, bigger seeds or shatterproof stalks have been deliberately picked out for cultivation and selective breeding, creating the domesticated varieties we eat today. Agriculture has always been an unnatural business.

As we saw in Chapter 5, Hermann Muller realized in the 1920s that an understanding of genetics might be used to accelerate and direct this process. By bombarding crops with radiation, it was possible to induce hundreds of mutations, some of which would produce new strains with useful properties that might never have evolved naturally.

Then, in the 1970s, came a still more powerful tool: recombinant DNA technology, which allowed new genes to be spliced into organisms. Plant breeding no longer had to rely on a 'hit-and-hope' strategy of inducing mutations and picking the ones that looked promising. Genes that confer desirable traits could be deliberately inserted into crops instead, using either a bacterial vector or a 'gene gun' that fires new DNA into the genome on tiny particles of gold.

The potential The first GM plant developed with this technology arrived in 1985 – tobacco engineered with a gene from *Bacillus thuringiensis* (Bt). This bacterium is toxic to many insects, and is used by organic farmers as a pesticide. The Bt tobacco made this insecticide itself, reducing

timeline

1927
Hermann Muller suggests the idea of genetic engineering

1985
Production of first GM crop, a tobacco plant engineered with a bacterial insecticide gene

the need for chemicals to be sprayed for pest control. Food crops took a little longer to develop, but the first – the Flavr Savr tomato, which had a longer shelf-life – hit the US market in 1994. A similar product was launched in Europe in 1996, and tomato puree that was clearly labelled as 'genetically altered' began to sell well.

More GM crops were soon rolled out by biotechnology companies such as Monsanto: the first wave included Bt varieties of cotton and soya beans, and maize and oilseed rape resistant to herbicides. Both the industry and publicly funded scientists began to talk up the technology's potential for addressing food shortages and malnutrition in the developing world, with GM crops that can tolerate salty soil or drought, or that produce higher yields.

One of the most exciting prospects is golden rice, developed in 2000 by German scientist Ingo Potrykus. This is enhanced with a daffodil gene, which makes it produce the precursor of vitamin A. Up to 2 million people die each year, and 500,000 go blind, because their diets are deficient in this essential nutrient. As many live in countries where rice is the staple food, the technology offers a simple route to better health.

The backlash GM crops have been embraced by farmers in many parts of the world. More than 100 million hectares are in cultivation, mainly in North and South America but increasingly in China, India and South Africa. More than half the world's soya is GM, and an estimated 75 per cent of processed foods sold in the US contain GM material.

The great exception, however, has been Europe. While the technology was accepted by American consumers with little fuss, it was launched in Europe at an inopportune moment. In the mid-1990s, it emerged that several dozen Britons had contracted a new form of Creutzfeld–Jakob disease, a fatal brain condition, from eating beef infected with mad cow disease or BSE – despite government assurances that there was no risk. This provoked a crisis of confidence in food safety, to which GM crops fell victim.

late 1990s
Backlash against GM crops in Europe begins

2003
UK government report declares no evidence of food safety problems, and calls for case-by-case assessments

UK field trials of three herbicide-tolerant crops suggest potential damage to biodiversity

2008
114 million hectares of GM crops grown worldwide in 23 countries. None are in the UK

Though there has never been much evidence that genetic engineering raises any special safety issues, groups such as Greenpeace whipped up a wave of public hostility to these new 'Frankenfoods'. Scientists were accused of meddling with nature – and just as feeding animal carcasses to cows had caused BSE, it was felt this might have unpredictable consequences for human health.

Other concerns centred on environmental impacts. In theory, herbicide-tolerant varieties should be good for biodiversity, by reducing the need for chemical sprays, but many green activists worried that the reverse might happen in practice. If farmers knew they could douse weedkiller with impunity, what would stop them from using more of it?

Further fears emerged from a study that suggested the Bt toxin made by many GM crops could kill non-target insects, such as monarch butterflies (see box). Organic farmers began to complain that GM pollen would contaminate their fields. Anti-GM activists started uprooting trial crops, and the public mood turned sour. Supermarkets cleared products from their shelves. While there has been no formal ban, no GM crops are grown in the UK at the time of writing, and just a single variety is licensed for commercial cultivation in the European Union.

Do GM crops kill butterflies?

In 1999, a study at Cornell University suggested that GM crops might threaten one of America's most iconic insects, the monarch butterfly. When monarchs ate pollen from Bt maize in the laboratory, 44 per cent died within four days. Here, campaigners said, was *prima facie* evidence of the ecological damage this technology could cause.

The threat, however, was exaggerated. Bt is toxic to monarchs, but they eat milkweed in the wild, not maize, and field studies show that the amount of Bt pollen that reaches milkweed is harmless. Though hundreds of thousands of hectares of Bt crops have been planted across North America, the monarch continues to thrive.

Case by case Some of these concerns are more legitimate than others. Food safety is probably a red herring – GM products have been eaten by millions of American consumers for over a decade without any adverse consequences, and the few studies that have suggested problems have been flawed. Environmental objections, however, may have more merit: the British field trials that were not ripped up found that herbicide-tolerant crops can hit biodiversity, depending on the spraying protocols that are used.

What this highlights is the folly of trying to consider GM crops as a whole. That a plant is genetically engineered says nothing about whether it can be safely eaten, or its likely impact on the environment. What matters is what the genes that are inserted do, and how the crop is then managed by the farmer. Some transgenic crops are likely to have ecological benefits when properly used, or to improve yields or produce more nutritious foods. Others may pose hazards, to the environment or to health. The technology has huge potential, but it is not a panacea. The only reasonable way to judge it is by looking case-by-case at the crops it is used to create.

Food safety

Perhaps the biggest GM food safety scare blew up in 1998, when Arpad Pusztai, of the Rowett Research Institute, claimed to have found that potatoes modified with an insecticide called lectin were harmful to rats. The work received huge publicity, but Britain's Royal Society highlighted serious flaws in the research – such as failure to use an appropriate control group. The finding is widely considered unreliable.

Another controversy surrounded the addition of a Brazil nut gene to GM soya, which inadvertently led to the transfer of a nut allergen, too. The problem, though, was identified before the crop was marketed, and it was withdrawn. While this GM variety could have been harmful, the case demonstrates the thoroughness of safety testing, and says little about the technology as a whole.

the condensed idea
Every GM crop is different

33 GM animals

Goran Hansson, 2007 Nobel committee: 'It is difficult to imagine contemporary medical research without the use of gene-targeted models. The ability to generate predictable designer mutations in mouse genes has led to penetrating new insights into development, immunology, neurobiology, physiology and metabolism.'

Fearless Mouse, OncoMouse, Mighty Mouse and Frantic Mouse sound like they could be members of a murine version of the *Teenage Mutant Ninja Turtles*. They are actually genetically modified animals that have transformed the way scientists investigate disease.

Since Rudolf Jaenisch, of the Massachusetts Institute of Technology, first injected foreign DNA into a mouse embryo in 1974, millions upon millions of GM rodents have been bred for medical research. They have been used to develop new therapies for conditions such as breast and prostate cancer, and they will be pivotal to translating recent insights into how genes influence disease into the next generation of drugs and vaccines.

Genetic engineering is also being used to turn animals into biological factories, which 'pharm' medicines or other useful chemicals in their milk. It promises to address the shortage of organs for transplant, by altering pigs so their hearts and kidneys can be given to humans. It could provide more nutritious meat, and it may even hold out hope of defeating malaria (see box).

Of GM mice and men The overwhelming majority of the GM animals that have lived to date are rodents, and most of these are mice. In the UK, where animal experiments are meticulously recorded for welfare

timeline

1974
Rudolf Jaenisch (b.1942) creates first GM mouse

1988
Creation of Harvard OncoMouse, a transgenic model for cancer research

reasons, more than one in three of the 3.1 million procedures that take place annually involve GM mice. Some of these transgenic rodents, such as the OncoMouse first created in 1988, have had genes added by infecting embryos with a virus – the OncoMouse has one that makes it susceptible to cancer. Others are 'knockouts', in which a gene has been silenced so that scientists can examine its effects.

The first knockout mice were created in 1989, through the work of Martin Evans, Mario Cappechi and Oliver Smithies, who were rewarded in 2007 with the Nobel Prize for Medicine. Evans's contribution was the discovery of embryonic stem cells (of which we will learn more in Chapter 35), and the insight that these master cells can be used to seed mouse embryos with genetically modified tissue. Cappechi and Smithies independently developed a way of exploiting recombination, by which chromosomes swap DNA, to target particular genes and switch them off.

When the techniques were combined, it became possible to breed mice that lack almost any gene (while any gene can be knocked out, sometimes the effects are fatal). The first knockouts lacked a gene called HPRT, which in

GM mosquitoes

Malaria, which is transmitted to humans by mosquitoes, takes up to 2.7 million lives each year, chiefly in Africa. A team at Johns Hopkins University is determined to beat it using genetic engineering, and has developed a GM mosquito that carries a protein that makes it immune to infection with the malaria parasite.

As malaria impairs infected mosquitoes' ability to reproduce, the GM variety should have an adaptive advantage were it to be released into the wild. This means that over time, the resistant insects should supplant their natural relatives, and bring the spread of the parasite to a halt. This approach, however, is controversial with some environmental groups, as it would involve replacing a naturally occurring species with a genetically engineered variant. None have yet been released into the wild.

1989
Work of Martin Evans (b.1941), Mario Capecchi (b.1937) and Oliver Smithies (b.1925) leads to creation of first knockout mice

2000
Creation of 'spider-goats', engineered with a gene that produces silk in milk

2006
ATryn becomes first 'pharmed' drug approved for sale

people causes a rare condition called Lesch–Nyhan syndrome, and others soon followed for cystic fibrosis, cancer and a host of human diseases.

Geneticists who want to find out what a gene does can now knock it out in mice, and watch what happens. When the gene for the myostatin protein is silenced, the result is 'mighty mice' that grow unnaturally large muscles. Mice that lack another gene are 'fearless' and cuddle up to cats. Scientists can now order off-the-shelf mouse models of the human diseases that interest them, with which to investigate their progression or to test candidate drugs. Frantic, for example, is a knockout mouse prone to anxiety, while others are genetically susceptible to Alzheimer's, heart disease, Parkinson's and diabetes.

Animal pharm Spider silk is among the toughest fibres known to science, with a tensile strength five times greater than steel. This property makes it attractive to industry, for use in cables, sutures, artificial ligaments, even bulletproof vests, but it has a major drawback. Spiders make too little of it, and they are territorial carnivores that are impossible to farm. Genetic engineering has offered an ingenious solution, often nicknamed 'pharming'. A Canadian company called Nexia has inserted two spider genes into goats, which now secrete spider silk proteins in their milk. These can then be extracted in large quantities, and spun into fibre.

Is genetic engineering of animals cruel?

The process of genetic engineering itself poses no danger to animal welfare, but the genes that are knocked out or added can have deleterious effects, depending on what they are. There is no reason to think that 'pharmed' GM animals are likely to be any different from conventionally bred examples: the evidence from 'spider-goats' and omega-3 pigs has not suggested any problems. But many GM laboratory animals, most of which are mice, are created solely for the purpose of modelling a human disease, so a degree of suffering will often be involved. Some will also be used to test new drugs or surgical techniques. Two-thirds of all GM mice in Britain, however, are used to provide cells or to maintain breeding colonies, and are never subjected to experiments.

A similar approach has been adopted by GTC Biotherapeutics, an American company, to add human genes to goats that then produce a blood-clotting agent in their milk. In 2006, ATryn, the drug made in this way, became the first 'pharmed' medical product to be licensed for human use.

No products from GM animals have yet been approved for human consumption in Europe or the United States, but several are not far from the market. Harvard University scientists, for example, have added a gene from the nematode worm *Caenorhabditis elegans* to pigs, so that they produce omega-3 fatty acids. A diet rich in these nutrients is linked to improved brain function and a lower risk of heart disease, but they are normally found only in oily fish. The piglets that could turn bacon into a health food were accordingly named Salmon, Tuna and Trout. There is nothing to suggest that GM meat, milk or eggs might be hazardous to eat – though whether consumers will accept them is another matter.

Another exciting application of farmyard genetic engineering is the prospect of creating pigs with 'humanized' organs, which would not be rejected by the immune system if transplanted into people. Thousands of people die each year waiting for new kidneys, hearts or livers, and pig organs are just the right size for human use. GM animals could solve the shortage at a stroke.

Such 'xenotransplantation', however, may yet founder on another aspect of genetics. The pig genome is littered with the DNA of viruses that have written themselves into its genetic code over millions of years. While these porcine endogenous retroviruses (PERVs) do the animals no harm, some seem capable of infecting human cells in culture. There is no telling what effects these might have if transplanted into people. Genetics, though, may also supply a solution: scientists have identified the receptors by which PERVs enter cells, and it may eventually be possible to switch them off and limit any threat to health.

the condensed idea
GM animals save human lives

34 Evo-devo

Sean Carroll, California Institute of Technology: 'All complex animals – flies and flycatchers, butterflies and zebras and humans – share a common "tool kit" of "master genes" that govern the formation and patterning of their bodies.'

Seen through a microscope, the early embryos of all mammals look so similar as to be indistinguishable. Even trained eyes cannot tell whether a ball of a few cells will grow into a mouse, a cow or a human. All are formed in the same way, from the fusion of an egg and a sperm, each carrying half a genome's worth of chromosomes, and their development follows much the same pattern for the first weeks of life *in utero*.

From an evolutionary point of view, this is not so surprising. Humans and mice diverged only around 75 million years ago, and it makes sense that our early embryonic growth should proceed along similar pathways. People and fruit flies, however, are much more distantly related. We are vertebrates, they are not, and our last common ancestor – probably something known as a 'roundish flatworm' – has been dead for well over 500 million years.

Yet the new science of evolutionary developmental biology, or 'evo-devo' for short, has shown that on a genetic level, humans and flies look remarkably alike. Despite our manifold physiological differences, many of the genes that build our bodies are not just similar, but identical. The same passages of DNA determine the position of flies' compound eyes and humans' simple ones, and piece together the two species' body parts in the right order. They are universal software programs, which can run as happily on the hardware of *Drosophila melanogaster* as on *Homo sapiens*.

timeline

1859
Darwin publishes *On the Origin of Species*

1865
Mendel identifies laws of inheritance

The developmental-genetic toolkit Evo-devo combines genetics and embryology to determine the ancestral relationships between different organisms, and to establish how their DNA causes them to develop in their particular ways. It is the science of how genotype defines phenotype.

Its critical insights began to emerge in the early 1980s, when two German-based scientists, Janni Nüsslein-Volhard and Eric Wieschaus, used chemicals to generate random mutations in flies, then bred them and followed their offspring's development from embryos into adults. When a mutation had an interesting effect, such as giving an insect extra wings, or legs on its head, the scientists traced it back to the gene responsible. This

Naming genes

While there is now a broad set of rules for naming genes, the scientists who made new genetic discoveries were long allowed to get away with calling them whatever they liked. Genetics is consequently rich in fanciful vocabulary. One of the first genes of the developmental toolkit to be identified is called hedgehog, because fruit fly larvae that lack a working copy are short and hairy, and look a little like the spiny animals. Mammals have a related gene that was named Sonic hedgehog, after the video game character, and fish have one called tiggywinkle, after Beatrix Potter's prickly heroine.

The fruit fly has a mutation named Cleopatra, because it is lethal in combination with a gene called asp. Another mutation is Ken and Barbie: like the dolls, the flies that carry it have no external genitalia. Many key genes found by Nüsslein-Volhard and Wieschaus have kept their German names, such as *kruppel* (cripple), and *gurken* (cucumber). Sometimes, even creative minds get stuck. The gene called ring has nothing to do with shape or function: it stands for Really Interesting New Gene.

early 20th century
Modern evolutionary synthesis developed

1980s
Discovery of Hox genes that lay out body plans

2001
Human Genome Project reveals that only about 2 per cent of the genome contains protein-coding genes

allowed them to pinpoint the function of dozens of genes, and the points at which they tell the embryo to develop in certain ways.

A set of just 15 genes turned out to govern the layout of the early embryo. These include clusters that sit in a line on the same chromosome, called Hox genes (Hox is short for the homeobox, a 180-letter stretch of DNA that they all share). These set down the shape of the fly embryo, giving it a front and back, segments and sides, and they are placed on the chromosome in the same order in which they sculpt the body from head to abdomen. Hox genes tell the head to grow antennae, and the thorax to grow wings and legs. When they are mutated, they result in monsters, such as flies with legs where the antennae should be.

Though mice (and humans) have more Hox genes than flies, they do exactly the same job, directing the formation of body parts in the order in which they occur on chromosomes. They are the key elements of a 'developmental-genetic toolkit' with which embryos are pulled into shape. These genes are so similar across species separated by hundreds of millions of years of evolution that it is even possible to transplant them from one animal to another, without losing function. Knock out a Hox gene in a fly, and replace it with the same gene from a mouse, and it is often impossible to tell that anything has happened. Human Hox genes do the trick just as well.

The Hox genes are just the most basic tools by which the body is made. Scores of others have been identified, which all perform a similar task in different species. The eyeless gene, for instance, is so called because flies that lack it grow no eyes. Knock it out, and replace it with its mouse equivalent, and the fly will grow perfectly normal eyes. This is particularly remarkable because insects have compound eyes, and mammals simple ones. The gene seems to say: 'grow an eye of the type you would normally grow', with other genetic instructions specifying which type is appropriate to the species.

'We didn't know it at the time, but we found out everything in life is so similar, that the same genes that work in flies are the ones that work in humans.'

Eric Wieschaus

Genetic switches This understanding that the shape of very different species, with radically different body plans, is governed by a fairly small fundamental set of body-building genes, raises an obvious question. If we share these genes with flies and mice, why do we not have wings, antennae and segments, or whiskers and tails?

The answer seems to lie in 'genetic switches' that turn genes on and off. Some of these are proteins known as transcription factors: these bind to sequences called promoters and enhancers, which surround genes and raise or lower their activity. Others are controlled by the 98 per cent of the genome that is not involved in making proteins – the stretches of apparent nonsense often known as junk DNA. Much of this seems to play a critical role in telling genes when they should be active, and when they should keep quiet.

What the Hox genes, and the other tools in the kit, achieve is to set networks of these switches in motion in particular cells, according to their positions in the body. These networks, in turn, determine which genes go to work and which remain dormant. Every liver cell, pancreatic islet cell and dopamine neuron carries the same basic package of genetic software. Specialized programs from this software package, however, are activated in each cell type.

These changing patterns of gene expression also explain how the same genes can achieve such different results in different organisms. Species diversity is greatly influenced by the way the same genes are used in idiosyncratic ways.

This helps to solve the riddle of how so few human genes – about 21,500, as our genome sequence revealed – are sufficient to build such a sophisticated organism. The wonderful complexity of the human animal emerges only in part from genes that carry instructions for making proteins that are unique to our species. Evo-devo tells us that the intricate network of switches that conduct this genetic orchestra is just as important, if not more so.

the condensed idea
Genes build bodies and cells

35 Stem cells

Christopher Reeve (d.2004), paralysed actor and embryonic stem cell research advocate: 'Embryonic stem cells . . . are in effect, a human self-repair kit.'

In ancient Gaelic legend, Tir na Nog was the land of eternal youth, where sickness, ageing and death did not exist. Ian Chalmers, of the University of Edinburgh, a Scot proud of his Celtic heritage, remembered the story in 2003, when he identified a gene with remarkable properties.

It is switched on only in the cells of early embryos, and it appears to be critical to their ability both to copy themselves indefinitely as if forever young, and to develop into any of the 220 or more cell types in the adult body. Chalmers named it Nanog, and it is one of the genetic keys to the unique properties of embryonic stem (ES) cells.

ES cells are the body's master cells, the raw material from which bones and brains, livers and lungs are grown. They occur only in young embryos, in which cells have yet to differentiate into the specialized tissues of the adult body. Because they are 'pluripotent', with the ability to give rise to any of these tissues, they have great medical potential. They could generate replacements for cells that become diseased or damaged in conditions such as diabetes, Parkinson's and spinal paralysis, but they are also a source of controversy. As they must be harvested from embryos, some religious groups consider their use unethical.

The stem cell controversy ES cells were first isolated from mice in 1981, by a Cambridge University team led by Martin Evans. Almost two decades later, in 1998, a group led by Jamie Thomson, of the University of Wisconsin, isolated human ES cells, raising hope that their versatility might be harnessed to treat diseases. If ES cells could be grown into

timeline

1981
Embryonic stem cells isolated in mice by Martin Evans

1998
ES cells isolated in humans by Jamie Thomson (b.1958)

dopamine neurons, which are lost in Parkinson's disease, they might be transplanted to treat it. For diabetes, they might be used to grow new beta cells that produce insulin.

Research into these cells has usually relied on embryos left over after in-vitro fertilization, though embryos are sometimes created specifically for this purpose. Such experiments have revealed how to grow these cells into self-perpetuating colonies or 'lines', often using a bed of mouse cells to provide critical nutrients (though this technique is being phased out). Scientists are now investigating which genetic and chemical cues make ES cells pluripotent, and then tell them to develop into specialized tissue.

The work has been opposed by people who consider it wrong to destroy embryos for any reason, including research into life-saving therapies. Most, but not all, of these critics base their objections on religious belief, and

Adult stem cells

Stem cells are not unique to embryos. Certain kinds of stem cells can also be found in the tissue of foetuses, children and adults, serving as a seed stock from which cells can be replenished and organs repaired. Bone marrow is especially rich in stem cells, as is blood from the umbilical cord.

As they do not require the destruction of embryos, research and therapy using adult stem cells are not controversial. They are already employed in treatments such as bone marrow transplants, and other applications have started clinical trials. Adult stem cells, however, are not as versatile as ES cells, as they have already taken steps towards differentiating into specialized tissue. Consequently, they may not be as useful for treating some conditions. Most scientists believe this promising branch of medical research should proceed in parallel with ES cell studies, not instead of them.

2006
Creation of induced pluripotent stem cells in mice by Shinya Yamanaka (b.1947)

2007
Creation of induced pluripotent stem cells in humans by Yamanaka and Thomson

also oppose abortion. Different countries have taken radically different approaches to the issue. The UK, China, Japan, India and Singapore are among the field's enthusiastic backers, both permitting ES cell research and supporting it with public money. Others, such as Germany and Italy, have banned or partially banned it.

The question has become most politically charged in the United States, the world's strongest scientific superpower, but also a nation with an influential religious right. In 2001, President Bush announced that federal funds could be used only to study ES cell lines already in existence, a compromise that satisfied few. The embryo-rights lobby still sees all research as immoral. Scientists and patient groups regard the rules as unnecessarily restrictive, and point out that as existing lines were grown with mouse cells, they will be unsuitable for transplantation. Several states, notably California, have set up their own funds for ES cell research, and private companies continue to invest in it.

Stem cells and cancer

One consequence of stem cells' ability to divide unchecked, and to differentiate into every tissue type, is that they can trigger grotesque cancers known as teratomas – the term means 'monstrous tumour' in Greek. These generally develop in foetuses, though they are often undiagnosed until later in life, and they can contain teeth, hair, bone, and even complex organs such as eyeballs or hands. One way in which scientists test the pluripotency of ES cells is to implant them into mice, to see if they generate teratomas. These cells' cancerous potential is a safety hurdle that must be cleared before they can be used in treatment. Many other cancers, such as acute lymphoblastic leukaemia, are also driven by cancer stem cells, which enable tumours to grow and spread.

The road to therapy No ES cells have yet been used to treat patients with any medical condition, though at the time of writing, Geron, a US company, is poised to start clinical trials. The cells, however, have been differentiated into a wide range of tissue types in the laboratory, which have been used successfully to treat conditions such as Parkinson's, muscular dystrophy and paralysis in animals. Genetic discoveries have also helped scientists to create a new kind of pluripotent stem cell by reprogramming adult tissue, which could address some ethical objections to the technology.

As well as Nanog, several other genes that are expressed in a particular pattern in ES cells have been identified. These include genes called Oct-4, LIN28 and three gene 'families' known as Sox, Myc and Klf. By genetically modifying adult tissue, so that these genes are switched on, it is now possible to turn back the clock on skin cells, so they acquire the pluripotency of embryonic cells. This was first achieved in mice in 2006, by a Japanese team led by Shinya Yamanaka, of the University of Kyoto, and in 2007, both Yamanaka and Thomson repeated the feat in humans. These induced pluripotent stem (IPS) cells have already been used to treat sickle-cell anaemia in mice.

IPS cells could have several advantages over standard ES cells. They do not require human eggs or embryos, which are in short supply. As they could be grown from the patient to be treated, they would be genetically identical and unlikely to be rejected by the immune system. They can also be made without destroying human embryos.

These advantages, however, do not make ES cell research obsolete. First, the techniques currently used to make IPS cells are too dangerous for therapy. The genetic modification is done with a virus that can promote cancer, as can one of the genes that is altered, called c-Myc. These cells also resolve ethical objections only in part. As Yamanaka and Thomson point out, they would not exist had scientists not been allowed to examine the genetics of ES cells in the first place.

The study of IPS cells is still in its infancy, and it is not yet known whether they behave in exactly the same way as ES cells. Stem cell scientists consider it critical to examine how both kinds of cells work, side by side. One kind may be better for some indications, and the other for others. It is too early to tell.

the condensed idea
Genes make master cells

36 Cloning

Ian Wilmut: The potential of cloning to alleviate suffering . . . is so great in the medium term that I believe it would be immoral not to clone human embryos for this purpose.'

The most celebrated sheep of all time, Dolly, was born in a Scottish laboratory on 5 July 1996. The ewe, created by Keith Campbell and Ian Wilmut, of the Roslin Institute in Edinburgh, was the first mammal to be cloned from an adult cell – a genetic copy of a living animal. As the cloned DNA was taken from a mammary gland, they named her after a famously ample-chested country-and-western singer: Dolly Parton.

Frogs and fish had been cloned decades before Dolly's birth, and in the 1980s, Russian scientists had cloned a mouse called Masha, by moving the nucleus from an embryonic stem (ES) cell into an empty egg. Every effort to make a mammalian embryo bearing an adult's DNA, however, had failed. In mammals, certain genes that are essential to embryonic development are always switched off in adult somatic cells, by a process called methylation, and this seemed to make cloning impossible.

Campbell and Wilmut, however, proved otherwise. They took the nucleus from a somatic (adult) cell of a sheep and placed it into an egg from which the nucleus had been removed, then coaxed it into dividing with electrical stimulation. Somehow, and the precise reasons remain unclear, this method can reprogram the nucleus and undo methylation, allowing a cloned embryo to grow. Dolly shared all her nuclear DNA with her somatic cell donor. Only the DNA in her mitochondria came from the ewe that supplied the egg.

The technique, known as somatic cell nuclear transfer (SCNT), was not efficient – it took the Roslin scientists 277 attempts to make Dolly. But by

timeline

1952
First cloned frog

1986
Mouse cloned using the nucleus of an embryonic stem cell

showing that it can work, they opened exciting possibilities. Prize livestock might be cloned in agricultural breeding programmes. And if SCNT worked for human cells, there could be medical applications.

Therapeutic cloning ES cells can grow into any tissue in the body, and could thus provide replacements for diseased or damaged cells. SCNT suggested that 'therapeutic cloning' might further enhance their medical utility. If stem cells were grown from an embryo cloned from the patient in need of treatment, these would share that patient's genetic code. The cells could be transplanted without fear of rejection by the immune system.

The technique could also create models of disease. DNA from patients with conditions such as motor neuron disease could be used to make cloned ES cells, carrying genetic defects that influence the disorder. Such cells would be valuable for studying the condition and testing new drugs.

Such developments, however, first required human embryos to be cloned by SCNT – a task that faced two major hurdles, one ethical and one technical. Even some people who approve of ES cell research object to therapeutic cloning as it could advance efforts to clone a human baby – a

Cloned food

The potential of cloning is not limited to medicine – its first benefits are more likely to be for animal husbandry. SCNT can be used to copy prize livestock with exceptional milk yields or musculature, to preserve genetic profiles that make them valuable to farmers. Clones would not be slaughtered themselves – they are too expensive – but used as breeding stock.

Food safety agencies in the US and the European Union have declared there are no scientific reasons why food from cloned animals and their offspring should not be safe for human consumption. The main objection is founded on animal welfare – cloning is still inefficient, and many clones have congenital abnormalities. But it is a safe bet that their meat and milk will reach the table soon.

1996
Creation of Dolly the sheep by Ian Wilmut (b.1944) and Keith Campbell (b.1954)

2004
Woo-Suk Hwang (b.1953) claims creation of first cloned human embryo

2005
Hwang's research discredited as fraud, but a UK team successfully clones a human embryo

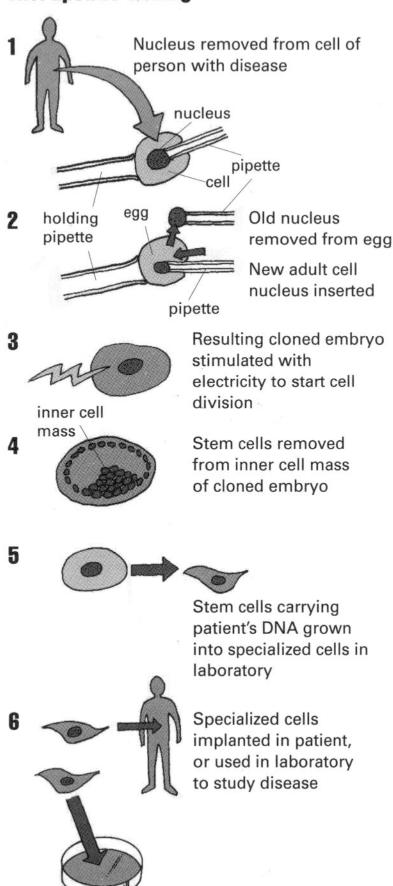

story told in the next chapter. And more critically, though SCNT was soon used to clone mice, pigs, cattle and cats, it is much more difficult to accomplish in primates.

The Hwang affair Countries that allow ES cell research have generally decided that the medical potential of therapeutic cloning outweighs the risks, and have permitted this application of SCNT while banning it for reproductive purposes. In February 2004, scientists from one of these nations, South Korea, claimed to have mastered the technical challenges involved.

In a paper published in the journal *Science*, a team led by Woo-Suk Hwang reported the creation of the world's first cloned human embryo and the extraction of ES cells. The following May, Hwang announced a still greater feat – the production of 11 cloned ES cell lines, each genetically matched to a different patient. Just as importantly, the group said it had refined SCNT so that fewer than 20 eggs were needed to generate a colony of cloned cells. With such a success rate, the technique could become medically viable.

It was all too good to be true. In November 2005, it emerged that Hwang had obtained eggs for his research unethically, and as his work fell under greater scientific scrutiny, it unravelled. His cloned stem cells had been faked: genetic tests proved that they had not been cloned at all. Of his purported achievements, only the creation of Snuppy, the first cloned dog, withstood independent genetic analysis. He had perpetrated a great scientific fraud.

Human embryos can be cloned, even if Hwang may not have managed it. In 2005, a team at Newcastle University succeeded, as have two American companies. None of these groups has yet made cloned stem cells, but such cells have been extracted from cloned monkey embryos. This ultimate goal is within reach.

Therapeutic cloning, however, has lost some of its lustre since the Hwang affair. Human eggs, which are essential, are always likely to be in short supply as they cannot be donated without risk. This means that even if cells can be cloned from patients, they are likely to be prohibitively expensive. Scientists are thus looking elsewhere for alternatives. For therapy, many think it will be more practical to use induced pluripotent stem cells, made by reprogramming adult tissue, or banks of ordinary ES cells from which suitable matches can be made to patients.

In research, another approach is to use SCNT to put human nuclei into empty animal eggs, to make 'cytoplasmic hybrids' carrying genetic material that is 99.9 per cent human. While these would be unsuitable for therapy, they could generate cell models of diseases, which British and Chinese scientists are already trying to create. The SCNT technique that created Dolly may never be used to create cloned cells for transplanting into patients, but it could still be a medical tool of great value.

'The amount of time and money needed to create these uniquely cloned solutions makes it unlikely that SCNT will provide a practical, widespread solution.'

Ruth Faden, stem cell scientist

Jurassic Park

In the 1993 movie Jurassic Park, dinosaurs are brought back from extinction by cloning, using DNA from mosquitoes preserved in amber, that had fed on their blood. While this makes great science fiction, most scientists think it impossible in practice. DNA from creatures that lived tens of millions of years ago will almost certainly be too degraded to use for cloning. On top of that, these animals have no living relatives sufficiently close to provide eggs for injection with dinosaur DNA.

Cloning, however, might be used to revive creatures that have become extinct more recently. In Australia, a project is underway to clone the Tasmanian tiger, using DNA from the last known animal, which died in 1936. It might also be possible to re-create the mammoth: an exceptional specimen recovered from the Siberian permafrost is thought to contain DNA of sufficient quality to attempt cloning. Its modern relative, the elephant, could plausibly act as egg donor and surrogate mother.

the condensed idea
Clones are genetic copies

37 Cloning people

Lord May, President of the Royal Society: 'Few disagree that it would be extremely irresponsible to try such an unsafe technology on people. It is important, therefore, that every country introduces effective legislation to deter cowboy cloners.'

The period between Christmas and New Year is always a quiet time for the media. In 2002, an obscure UFO cult found a way to fill it. The Raelians, a group founded by a French motor-racing journalist who believes humans were created by aliens, called a press conference on 27 December to announce the birth of a girl called Eve. The baby, they claimed, was the first cloned human being.

This seasonal nativity tale captured headlines around the world, though to experts, it was a transparent hoax. At the time, even respected researchers had failed to clone a human embryo, let alone a living baby, and the Raelians had never cloned so much as a frog. They produced no evidence that Eve existed, never mind presenting her for genetic tests that might prove their claim. The cult had launched a company offering a $200,000 cloning service to couples seeking to recreate dead children. The story seemed to be a cynical publicity stunt.

Nevertheless, it provoked outrage as well as disbelief. Scientists pointed out that while cloning sometimes worked in animals, it was highly inefficient, causing dozens of miscarriages and deformities for every live birth. To attempt it for human reproduction would be unethical in the extreme.

Even were it safe, the very idea of cloning people caused widespread disgust. To force somebody to share the DNA of another person struck many as an insult to human dignity. 'The cloned individual', said Leon

timeline

1986
First mouse cloned, from embryonic stem cell

1996
Birth of Dolly the sheep, the first mammal cloned from an adult cell

Kass, President Bush's bioethics adviser, 'will be saddled with a genotype that has already lived.' The vanity of a wealthy megalomaniac, or the misguided grief of a bereaved parent, would create a life to be lived in another's shadow.

The Raelians appeared to be cranks, but they were not alone. Two maverick fertility doctors, Severino Antinori and Panayiotis Zavos, claimed to be pursuing similar work. Their efforts led most governments to outlaw reproductive cloning, and the United Nations is pursuing a global ban.

Cloning in fiction

Reproductive cloning is a staple of science fiction, which has perpetuated the mistaken belief that clones would be identical to their DNA donors in every respect. In general, movie clones and clonees are played by the same actor, when in practice there would be no guarantee of more than a family resemblance. Arnold Schwarzenegger plays his character and its clone in *The 6th Day*, as do Ewan McGregor and Scarlett Johansson in *The Island*, and Michael Keaton in *Multiplicity*. In *Star Wars: Attack of the Clones*, Temuera Morrison does even better, playing both the bounty hunter Jango Fett and an entire army cloned from his DNA.

What would a human clone be like? In the absence of a sudden torrent of evidence from the Raelians, Zavos or Antinori – and few scientists are holding their breath – it is safe to assume that no human clone has yet been born. The feat, though, may not be impossible, and it could yet be achieved in a country that is unwilling or unable to prevent it. What, if that were to happen, might such a cloned person be like?

For a start, it would probably be deformed, if not stillborn. Animal cloning has improved since it took 277 eggs to create Dolly the sheep, but it is still fraught with technical difficulties, particularly in primates. The somatic cell nuclear transfer process seems only sometimes to reset adult DNA

2001
UK bans human reproductive cloning, while permitting therapeutic cloning

2002
Raelian cult claim of human cloning dismissed as hoax

correctly, and clones of all species suffer a high incidence of the imprinting disorders described in Chapter 29. Many are born oversized, or with heart and lung defects. Those that survive infancy often die young – Dolly herself was put down at six, half the usual life span for a sheep, after developing lung disease (though it is unknown whether this was related to cloning). Many also have shortened telomeres – structures at the end of chromosomes that protect them against DNA damage – a symptom of premature ageing. Human clones could be expected to suffer from all these problems. The cost in lost pregnancies and dead and deformed children explains why almost every scientist thinks reproductive cloning is currently unethical.

A human clone would share all the nuclear DNA of the parent from which it was created. That does not necessarily mean, however, that it would be a carbon copy, with similar looks, abilities and personality. While genetics certainly influences these traits, it does not determine them as if laid down by a blueprint. Identical twins share all their DNA, yet while they tend to be more similar than fraternal twins, they are by no means the same.

Clones would actually be more different from their donors than identical twins are from one another, since they would not share the same womb, or a similar childhood environment, family and peer group. As John Harris, a philosopher of bioethics, says: 'Since we know that all these experiences affect the structure of the brain, there is no significant sense in which any clone should be determined to be like its genome donor.' The widely held view that cloning could be used to resurrect Hitler, as in the movie *The Boys from Brazil*, or to replace a dead child, is a fundamental misunderstanding. Copying genes does not mean copying a person.

Is reproductive cloning wrong? While there is widespread disagreement over the morality of therapeutic cloning, it is difficult to find people of any ethical persuasion who think reproductive cloning is acceptable today. The safety issues involved are too great. But such considerations are contingent on technology, and might plausibly be overcome. That raises an interesting thought experiment. If animal research suggested reproductive cloning could be safe, some people might want to try it – perhaps couples who are infertile because the man makes no sperm. Would it be intrinsically wrong for them to do so?

Mitochondrial transfer

A form of nuclear transfer that is subtly different from cloning is currently being investigated to allow women with diseases caused by faulty mitochondria to have healthy children. Mitochondria are small structures outside the nucleus that generate cells' energy, which we all inherit from our mothers. They contain a few genes, and mutations can cause kidney, brain and liver disorders that mothers will transmit to their offspring.

To prevent this, a team at Newcastle University is developing a way of transferring the nucleus from an affected woman's egg to a donated egg with healthy mitochondria, from which the nucleus has been removed. This would then be fertilized with sperm from the patient's partner. The technique is a little controversial because a child born this way would have DNA from three parents. The nuclear DNA would come from its mother and father, but its mitochondrial DNA would come from the egg donor.

Though most people instinctively recoil at the prospect, it is far from clear that the answer is no. Cloning is unnatural, but so are artificial insemination, in-vitro fertilization and, indeed, the entire practice of medicine. Clones would share their DNA with other people, but so do identical twins, who suffer no loss of individuality or dignity. They would face discrimination and stigma; but so, not so long ago, did children born out of wedlock.

Human reproductive cloning may turn out to be impossible, or impossible to attempt without intolerable risk. It cannot replicate people, and it will never appeal to more than a tiny minority – alternative breeding options will remain more reliable and cheaper. Today, it is the domain of charlatans and cowboys. It may not be for ever.

the condensed idea
Clones are not carbon copies

38 Gene therapy

Len Seymour, British Society for Gene Therapy: 'For patients without bone marrow donors, gene therapy now provides a remarkable, potentially curative, way forward.'

Ashanti DeSilva is an American college student in her early 20s. Yet when she was born in 1986, she was not expected to attend high school, let alone university. Ashanti suffered from a rare recessive disorder called severe combined immune deficiency (SCID). It meant she lacked a functioning immune system, leaving her dangerously exposed to each and every passing germ.

Children with SCID live forever on the edge of disaster. As they cannot fight off pathogens, even mild infections can be life-threatening. Many die in infancy, and those who survive are often shielded from the outside world in a sterile pouch – the condition is often known as 'bubble baby' syndrome. They cannot attend school or mix with other children, and without a bone marrow transplant from a matching donor, few live to adulthood.

No suitable donor could be found for Ashanti, but in 1990, researchers at the National Institutes of Health came up with an alternative. A team led by French Anderson removed some of her ineffective white blood cells from her body, and infected them with a virus modified to carry a healthy copy of her faulty gene. When these treated cells were infused into her bloodstream, Ashanti's immune function improved by 40 per cent. She became well enough to go to school, and even to receive vaccines, which cannot normally be given to immune-compromised patients. She was the first patient to be successfully treated with gene therapy.

Our friend the virus Gene therapy did not cure Ashanti: the genetically modified cells worked only for a few months at a time, and she has had to return for regular treatment. For that reason, the technique was

timeline

1990
First successful use of gene therapy by French Anderson (b.1936)

1999
Death of Jesse Gelsinger (1981–99) during gene therapy trial

at first used only when a bone marrow transplant was not an option. In 2000, a team at Great Ormond Street Hospital in London, and the Necker Hospital in Paris, advanced the procedure further, to correct the SCID mutation in the bone marrow of children, which should offer an indefinite cure. Its early success raised hope that the strategy might be effective against this and other inherited diseases.

The therapy works by harnessing the aggressive properties of one of humanity's microscopic enemies. When viruses infect us, they reproduce by introducing their genetic material into our cells, hijacking their replication machinery and forcing it to churn out more viruses. One class, the retroviruses, actually write themselves into our genome with specialized enzymes.

Medicine can exploit this viral talent and turn these pathogens into vectors for transporting fresh DNA into cells. Virulence genes are excised to make the virus harmless, then a normal copy of the defective human

germline gene therapy

All the gene therapies attempted to date work on the somatic cells that make up the vast majority of the body's tissues and organs. They aim to correct genetic defects in an individual patient, but as they do not alter germ cells that produce eggs and sperm, those mutations can still be passed on to children.

Future technologies may go further, to create 'germline gene therapies' that change genes of both patients and their offspring. This is more controversial, because people who are yet to be born have no say over genetic manipulations that might have unforeseen consequences. Proponents of germline gene therapy, however, do not see what the fuss is about – at least where disorders such as SCID or cystic fibrosis are involved. If it is possible to root a deleterious gene out of a family for good, they ask, why would it be wrong to do so?

2000
Successful use of new gene therapy technique for SCID by Anglo-French team

2002
Anglo-French trial interrupted after several patients develop leukaemia, one of whom dies

2008
Successful use of gene therapy to treat Leber's congenital amaurosis, a genetic cause of blindness

gene that needs to be replaced is spliced into its genetic code. When a patient's cells are infected with this modified virus, they take up the new gene, and should start making the normal protein. The principle is the same as downloading a 'patch' to debug misfiring computer software.

With some viral vectors, such as adenoviruses that commonly cause tonsillitis, the new gene will be active only in those cells that were infected: when these die, their successors will not express the added trait. That is why Ashanti DeSilva needed repeated treatments. If a retrovirus is used, however, the new gene will be incorporated into the genome of infected cells, and passed on to its descendents. The genetic defect should be corrected for good.

Gene therapy

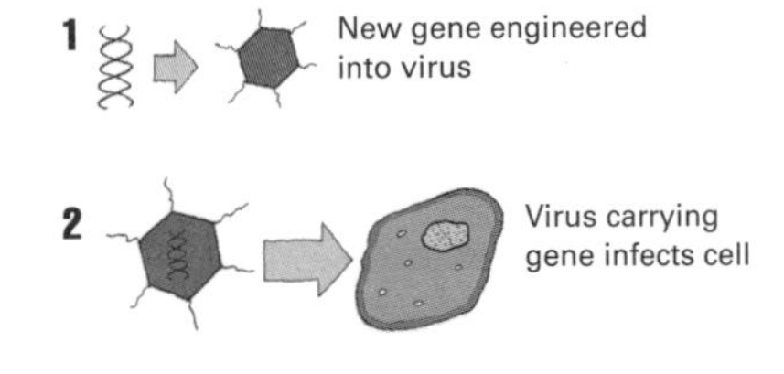

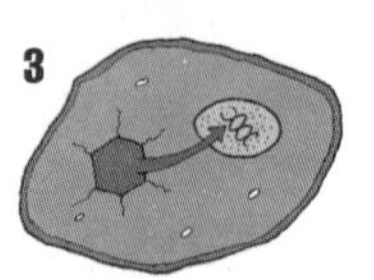

4 New gene begins making protein, to treat disease

Unintended consequences Viral vectors are critical to existing methods of gene therapy, but they are also its greatest weakness. They can affect the human body in unpredictable ways, causing side-effects that have greatly limited the technique. The Anglo-French SCID trial, which used a retrovirus, might have corrected the disorder, but this success has come at considerable cost. Five of the 25 children treated so far have developed leukaemia.

When a retrovirus incorporates itself into the genome of host cells, doctors cannot control the position at which it burrows its way in. Sometimes, it interrupts an oncogene, triggering uncontrolled cell division and cancer. As 80 per cent of children with leukaemia recover, while untreated SCID is invariably fatal, it can be argued that this risk is worth running. Only one of the affected children has died, while three are in remission, and one had been recently diagnosed at the time of writing. But it is by no means ideal for a therapeutic technique that has been claimed as a standard-bearer for genetic medicine.

Leukaemia is not the only unwelcome consequence that viral vectors can have. In 1999, Jesse Gelsinger, an 18-year-old with a genetic liver disorder, took part in a University of Pennsylvania trial of a gene therapy designed to treat his condition. He suffered a massive immune reaction to the adenovirus vector, which killed him. As well as being a personal tragedy, it was a severe setback to the field.

Gene doping

It is already hard enough to detect athletes who use performance-enhancing drugs such as human growth hormone. Gene therapy could make it even more difficult. Scientists have already used the technology to alter the genes of mice and monkeys, so that they produce increased amounts of proteins that boost strength or endurance, such as erythropoietin (EPO). Such 'gene doping' by athletes could be virtually impossible to prove. Those found to have excess EPO in their systems could claim, correctly, that their genes were responsible. Sophisticated genetic tests, not yet available, would be needed to show that genetic enhancement had taken place.

Adenoviruses and retroviruses are now being replaced in gene therapy trials by a different vector, the adeno-associated viruses. Unlike retroviruses, these always incorporate themselves in the genome in the same, safe place, and unlike adenoviruses, they do not normally cause human diseases, making immune overreactions unlikely. A trial based on this approach has improved the sight of four patients with Leber's congenital amaurosis, a single-gene cause of blindness. Non-viral vectors, such as synthetic zinc-finger proteins, are another promising option.

But while these kinds of gene therapy are likely to prove safer, and perhaps more effective, scientists are not as excited about this technology as they once were. Although it is promising for a few single-gene disorders, it has failed in many more. It is one thing to modify enclosed tissues, such as bone marrow and retinal cells, but quite another to correct genetic defects that have more systemic effects, such as the cystic fibrosis mutation.

Most diseases, moreover, are not caused by single genes, but are influenced by multiple genetic variants that each slightly raise risk. Diabetes can be affected by two dozen genes, and it will be impractical to alter them all. While gene therapy will have a place in the clinic, it is no panacea for inherited disease.

the condensed idea
Mutations can be corrected, sometimes

39 Genetic testing

Kari Stefansson, of deCODEme: 'If, as a competent adult, you choose to look at your risk of developing Alzheimer's, that is your prerogative. But no one will force you to look at your Alzheimer's risk if you do not want to.'

In the English city of Cambridge, a cycle path is decorated with more than 10,000 lines, each in one of four colours. The pattern follows the sequence of a gene on chromosome 13, which was identified in 1995. The gene is BRCA2, which takes its name from the disease that often results when it is defective: breast cancer.

One in nine women in developed countries will get breast cancer during their lifetimes. Among women who have mutations in the BRCA2 gene, however, up to four-fifths will develop the disease, and a similar risk applies to defects in another gene called BRCA1. Both are tumour suppressors, which normally stop cells from becoming cancerous. Women who are unlucky enough to inherit mutations lack a critical line of defence, which leaves them peculiarly vulnerable to cancers of the breasts and ovaries.

Thousands of women belong to families with a long history of breast cancer, who have often lost mothers, grandmothers, sisters and aunts. The isolation of the BRCA genes has meant that some of these people have been able to discover whether the family risk applies to them. If a BRCA mutation is known in a relative with breast cancer, women can be tested to determine whether they have inherited it too. For those who test negative, the procedure can offer peace of mind, while those with a positive result

1993
Identification of Huntington's disease mutation

1995
Identification of BRCA2 gene

Genetic matchmaking

Tay–Sachs disease is a recessive Mendelian disorder that causes neurological damage and death, usually in infancy. The allele that causes it is common among Ashkenazi Jews – possibly because carriers of one copy are partially protected against tuberculosis, an advantage in the ghettos in which Jews have often been forced to live.

In orthodox and conservative Jewish communities, it is common for matchmakers to arrange marriages, and many now use genetic testing as a tool. Young people are tested for their Tay–Sachs status, so that carriers of the damaging allele can be kept apart. Were two to marry, their children would have a one in four chance of developing the disease.

can take action to reduce their risk. Most have regular mammograms to ensure any incipient tumour is detected early, and some even opt for preventive mastectomies.

Testing dilemmas BRCA1 and BRCA2 are just two of the genetic diseases for which it is now possible to test. Newborn babies, for instance, have their heels pricked with a sterile needle a week after birth, to collect blood that is then screened for inherited disorders such as phenylketonuria (PKU). In the UK, about 250 infants test positive each year, and can thus be protected against the neurological damage that PKU would otherwise inflict.

Other reliable tests are available for hundreds of disorders caused by defective single genes. Often, as with PKU or haemophilia, their results ensure that patients receive appropriate treatment. Even for incurable diseases, such as cystic fibrosis and Duchenne muscular dystrophy, genetic diagnosis can help doctors to manage symptoms, and parents to prepare themselves for the future.

2001
Completion of first draft of Human Genome Project, at a cost of $4bn

2007
Launch of deCODEme and 23andMe consumer genotyping services

2008
Applied Biosystems sequences an individual's genome for $60,000

Some genetic tests, however, are more problematic. Huntington's disease is a prime example. As it is caused by a dominant mutation, anybody with an affected parent has a 50 per cent chance of having inherited it. Yet though there is a reliable test, many of those at risk refuse to have it – including Nancy Wexler, the scientist whose research led to the test's development (see Chapter 19). Huntington's is a late-onset condition that causes progressive cognitive decline, it is invariably fatal, and there is no cure. As a positive test is tantamount to a death sentence, many people would rather not know.

Another genetic dilemma is posed by amniocentesis, which can be used to test the developing foetus for abnormalities such as Down's syndrome. If the result is positive, nothing can be done. A couple must decide whether to progress with a pregnancy that will produce a disabled child, or to have an abortion.

Consumer tests All the genetic tests described so far belong to clinical medicine: they are available only through doctors, and after appropriate counselling. They look for rare but major mutations that always result in disease, or which dramatically raise risk. Most genetic influences on health, however, do not work that way: they involve common variations that slightly raise or lower people's chances of developing diabetes or cardiovascular disease. Tests for these variants pose new challenges, not least because they are increasingly being marketed directly to consumers.

Two companies that offer such services, deCODEme and 23andMe, were launched in 2007. For $1,000, they will take DNA from a mouth swab and scan a million single nucleotide polymorphisms (SNPs) – points at which the genetic code varies between individuals. The results are used to assess customers' risk of more than 20 diseases, as well as other aspects of inherited physiology such as male pattern baldness.

In theory, such information should be of great value to people's health, allowing them to change diet or lifestyle to counter inherited risk, or to ensure they get regular screening. But these tests can create problems, too. The variants that are examined are not like the BRCA genes – they have only a small impact on disease risk, and environmental factors matter too. Only a handful of SNPs that influence these diseases are yet known, so the results are necessarily going to be incomplete.

This means that personal genotyping can very easily mislead. There is a danger of providing false reassurance that promotes a cavalier attitude to health: people with SNPs that suggest a lower risk of lung cancer might become less likely to quit smoking. Equally, apparently frightening results may provoke needless anxiety, especially when people are tested by online services that offer no counselling or medical advice. If you have an allele such as ApoE e4, which raises the risk of Alzheimer's sixfold, would you really want to find out through a website? When James Watson's personal genome was sequenced (see box), he asked not to be told his results for this gene.

> **'This test could cause unnecessary worry about potential health risks, or give others a false sense of security.'**
>
> **Joanna Owens, Cancer Research UK**

Personal genetic testing, though, is only going to become more common as the costs come down. It took $4bn to sequence the entire human genome for the first time, but personal testing can now be done for as little as $100,000. Most scientists think the cost will fall to $1,000 or less within the next five years. That will open up exciting medical possibilities, but many of the clues revealed are going to be cryptic and devilishly difficult to interpret.

Personal genomics

When the human genome was first sequenced, the published results were averages, composed of data from several people. The falling cost of the technology has now allowed two individuals – Craig Venter and James Watson – to have their personal genomes decoded. Venter's genome, published in 2007, cost $10m, and Watson's, published a year later, cost just $1m. The price tag is coming down still further – in 2008, a company called Applied Biosystems mapped the genome of an anonymous Nigerian for $60,000.

The X Prize Foundation, which has already run a contest to launch the first private space flight, has now endowed a genomics award to stimulate further technological developments. The $10m prize will go to the first team to sequence 100 anonymous human genomes in 10 days, at a cost of no more than $10,000 apiece.

the condensed idea
DNA can warn or mislead

40 Tailor-made medicines

Paul Martin, Nottingham University: 'There's no commercial incentive for the big companies selling existing drugs to go to the expense and effort of developing a test that will ultimately narrow down the number of people who take their drug.'

Shortly before the first drafts of the human genome were published in 2001, Francis Collins, the head of the public sequencing consortium, set out his vision of our genetic future. By 2010, he predicted, science would understand how genes contributed to a dozen common illnesses, such as diabetes and heart disease, paving the way for preventive therapies. Give medical research another decade, and these conditions would be treatable with 'designer drugs', created using genetic insights and prescribed according to patients' genotypes. Roll the clock on to 2030, and genomic medicine would have extended average life spans to 90 in the developed world.

This sort of futurology might sound far-fetched, but its first elements are already being realized. As we saw in Chapters 20 and 21, genetics has informed the design of treatments for diseases as diverse as HIV, flu and cancer. Herceptin, which works only for breast tumours with a particular genetic profile, has saved lives. Genetic testing is starting to allow people to predict their risk of developing certain diseases. Collins' timetable does not look wrong yet.

Bespoke medicine One of the next advances may be bespoke medical treatment, also known as pharmacogenomics, by which drugs are tailored to patients' genes. At present, most medicines work – or often

timeline

1960
Identification of the Philadelphia chromosome as common cause of chronic myeloid leukaemia (CML)

1990s
Development of Gleevec for treating Philadelphia chromosome positive CML

don't – on a one-size-fits-all basis. Drugs are tested on randomly selected patients, and must show themselves to be safe and effective in large samples before they reach the market.

Pharmaceutical companies hope to hit on blockbusters that can be sold to millions of patients. The model examples are statins for high cholesterol, and selective serotonin reuptake inhibitors (SSRIs) – the class of antidepressants that includes Prozac. Each manufacturer usually offers a slightly different version, and doctors and patients often work their way through the options by trial and error before alighting on the one that seems most effective.

Pharmacogenomics promises to change this. The way in which different drugs are metabolized is influenced by genetic factors, and as scientists start to understand these it should be possible to start prescribing accordingly. Test results will predict that certain patients should respond best to certain drugs: a Massachusetts General Hospital study, for example, has suggested that patients with a particular genetic type may not benefit from one SSRI. Genetics could also indicate whether a higher or lower than usual dose might be needed. And such profiling should also make medicine safer, telling doctors which pharmaceuticals to avoid because a person's DNA puts them at risk of an adverse reaction.

Gleevec

Chronic myeloid leukaemia (CML) is a blood cancer caused by the unchecked growth of certain white blood cells. It is often triggered by a type of genetic mutation known as a translocation, in which parts of chromosomes 9 and 22 fuse to create an abnormal structure called the Philadelphia chromosome. This produces a mutant protein that turns cells cancerous.

Treatment of CML has been transformed since 2001 by a drug called Gleevec (Glivec in Europe) – a least in those patients whose condition is caused by the Philadelphia chromosome. It blocks the mutant protein's activity, so that white blood cells are no longer produced in uncontrolled fashion. It is one of the first successes for pharmacogenomics.

1998
Launch of Herceptin

2001
Human genome drafts completed

2007
UK health service makes Herceptin available for appropriate patients

A related approach will be to narrow down diagnostic categories, so that patients are no longer considered to have type 2 diabetes or colon cancer, but sub-types of those conditions that are influenced by particular genes. Every case of diabetes is unlikely to have the same molecular pathway at its core. There may be several combinations of genes that affect the condition, each of which operates in different ways and requires a different treatment strategy. Genetic testing should help doctors to select the right tool for every job.

This could be especially useful for conditions that are difficult to treat, such as autism and schizophrenia. These are both influenced by a plethora of genetic variations, their symptoms often differ between individuals, and they may not be single disorders at all. If genetics can refine diagnosis, better therapeutic strategies should follow.

A new economic model Pharmacogenomics thus has great potential for patients, promising medicines that are more likely to work for them. But it also makes the pharmaceutical industry nervous, as it challenges its traditional business model. If the next generation of drugs is going to be targeted at genetic niche groups, it is simply not going to be possible to sell them in vast numbers, as it is with statins or SSRIs. Yet

Nutrigenomics

Our genetic profiles can affect the way we respond to particular foods: people with the phenylketonuria mutation, for example, need to follow a special diet to avoid brain damage. It is also likely that common genetic variations influence our nutritional needs, and this has led companies to launch 'nutrigenomic' services that purport to generate genetically tailored diets.

Nutrigenomics may well have a future, but the links between genetics and nutrition remain so poorly understood that most scientists think it does not yet offer value for money. Some criticize existing services as 'health horoscopes', which is not a bad analogy. They generally dispense vague platitudes that are harmless enough, but which could apply to anybody. Most suggest eating more vegetables and less fat – sensible advice for sure, but sensible for everybody, regardless of our genes.

many of the costs of drug development are fixed, which has led many observers to conclude that bespoke medicines will be very expensive. Herceptin, one of the first examples, is a case in point. At an annual bill of £20,000 per patient, many NHS trusts in Britain initially refused to pay for it. They relented only in the face of government pressure, court cases and a media campaign.

Herceptin has been the first skirmish in what could prove to be a drawn-out battle. But it also points towards a new economic model for drug discovery, which suggests some fears over the costs of pharmacogenomics are exaggerated. If a drug like Herceptin is designed for people with a particular genetic profile, it can be tested on that patient group. There is less risk of staging an expensive trial that generates negative results – the biggest cost of pharmaceutical research.

‘The drugs that we give in 2020 will for the most part be those that were based on the understanding of the genome, and the things that we use today will be relegated to the dustbin.’

Francis Collins

Furthermore, drugs that are known to be highly effective for certain patients can essentially market themselves. Doctors know Herceptin is the best option for women with HER-2 positive breast cancer, and that Gleevec is best for chronic myeloid leukaemia – their benefits don’t need to be advertised. This subtracts another substantial cost from big companies’ balance sheet.

Bespoke prescribing also has the potential to ‘rescue’ drugs that perform poorly at a population level, but which work for individuals. Many medicines are abandoned because they fail trials, or adversely affect a minority of patients. If it is possible to identify groups in whom these are safe and effective, some of this investment might still be recouped. Public health services and insurers, too, will save money by no longer paying for the mass prescription of broad-spectrum drugs that are useless for many patients.

The advent of pharmacogenomics will probably require drugs companies, as well as doctors, to change the way they operate. That does not necessarily mean, however, that the costs of medicine have to be driven inexorably upwards.

the condensed idea
Drugs can be tailored to genes

41 Designer babies

Francis Collins, National Human Genome Research Institute: 'Those well-heeled couples who decide they want a virtuoso musician may be disappointed to discover that their son turns into a sullen adolescent who smokes marijuana and doesn't speak to them.'

Debbie Edwards once thought she would never have children. Her nephew had inherited a genetic condition called adrenoleukodystrophy, and a test had revealed that she carried the mutation responsible on one of her X chromosomes. As she was a woman, with a second X chromosome bearing a working copy of the gene, Mrs Edwards was perfectly healthy. Any son she might conceive, though, would have a 50 per cent chance of developing progressive brain damage and dying young. She took the painful decision not to try to start a family.

Yet on 15 July 1990, Mrs Edwards gave birth to twin girls, Natalie and Danielle, at Hammersmith Hospital in London. She had not had a change of heart about the danger of adrenoleukodystrophy: science had found a way to avert it. The development of an embryo-screening technique had allowed her to conceive, confident in the knowledge that she would not have sick children.

The Hammersmith team, led by Alan Handyside and Robert Winston, created embryos by in-vitro fertilization (IVF), and grew them in the laboratory until the eight-cell stage. One cell was then removed from each embryo, so the scientists could examine the sex chromosomes to determine which were male and which were female. As adrenoleukodystrophy is

1978

Birth of Louise Brown, first baby born by in-vitro fertilization

X-linked, and affects only boys, only female embryos were implanted into Mrs Edwards' womb.

The procedure is known as pre-implantation genetic diagnosis, or PGD, and Natalie and Danielle were the first examples of what the media called, a little misleadingly, 'designer babies'.

The PGD revolution PGD babies have not, in truth, been designed. Their DNA has not been altered, but the term has stuck because the technology allows parents to do something that has never before been possible. They can choose between potential children on the basis of genetic qualities that they do or don't have – much as a shopper might choose a designer dress.

The advance gave couples who knew they might pass on a devastating inherited disease the opportunity to have healthy

Pre-implantation genetic diagnosis

1 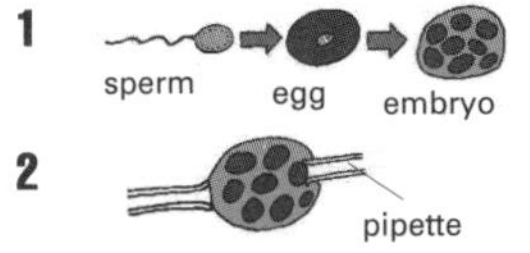

Embryos created by IVF

2 pipette 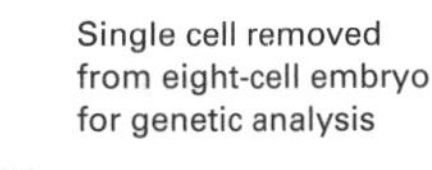

Single cell removed from eight-cell embryo for genetic analysis

3 ✗ ✗ ✓ ✓ ✗

Embryos without genetic defect selected for transfer to womb

4 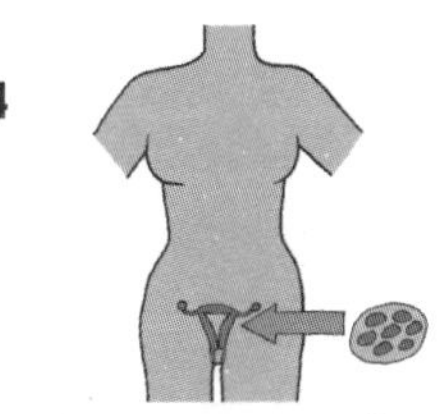

Disease-free embryo transferred to womb

Pre-implantation genetic screening

The embryo biopsy technique also has applications in fertility treatment, to check the genetic quality of embryos with a view to improving the chances of a successful pregnancy. Most embryos with too many or too few chromosomes will miscarry, and the test can be used to count them so that only normal ones are selected for transfer.

In the UK, eight clinics are licensed to perform this pre-implantation genetic screening, for women with a history of miscarriage or failed IVF, but there is controversy about whether it works. A 2007 Dutch study suggested it might actually reduce IVF success rates, probably because the biopsy can damage the embryo. The technique's advocates, however, point to methodological problems with that research, and argue that when properly performed, it has clear benefits for some women.

1990

Development of pre-implantation genetic diagnosis at Hammersmith Hospital in London, and birth of Natalie and Danielle Edwards

2002

Birth of Adam Nash, first child conceived as a 'saviour sibling'

offspring. At first, PGD was sensitive enough only to prevent X-linked disorders such as haemophilia or Duchenne muscular dystrophy, by testing for sex, but it soon became possible to screen for autosomal conditions such as cystic fibrosis and Huntington's as well. More than 200 diseases can now be detected, and several thousand PGD babies have been born worldwide.

The technique, however, has stirred ethical controversy. Those who object to the destruction of embryos think PGD immoral because those embryos that carry genetic mutations are thrown away or donated for medical research. Particularly contentious is its application to genes such as BRCA1. Mutations in this gene substantially raise the risk of breast cancer, but do not invariably cause it, and those who inherit them can protect themselves, albeit with mutilating prophylactic surgery. Critics see embryo screening as a form of eugenics, of rooting out the disease by eliminating the carrier.

PGD has also offered a lifeline to existing children who suffer from diseases such as leukaemias and anaemias, and need a cell transplant from a matching donor. If no such match exists, parents can try for another baby and screen their embryos to select those suitable to donate. In 2002, a young American girl named Molly Nash, who suffers from Fanconi anaemia, became the first to be successfully treated with tissue from such a 'saviour sibling'. She received stem cells from the umbilical cord of her newborn brother, Adam, whose tissue type had been screened as an embryo.

This extension of PGD has raised another concern. The biopsy process carries a very small risk to the embryo, and some people think it wrong to incur this when that

Artificial chromosomes

PGD is only capable of choosing between embryos with genetic profiles inherited from their parents, but true 'designer babies' might one day become possible through advanced genetic engineering. If this does happen, one route may be to use synthetic chromosomes, deliberately engineered to carry beneficial genes and inserted into the cells of early-stage embryos.

This approach probably remains decades away from plausibility, but it would have two advantages. It would not interrupt the genetic sequence of existing chromosomes, reducing the risk of introducing an error that might cause a disease such as cancer. And biophysicist Gregory Stock, a proponent of genetic enhancements, has suggested it might be possible to activate artificial chromosomes at a later date. Children who receive them could thus choose whether to switch on their genetic modifications on reaching adulthood.

embryo will not directly benefit. Britain's embryology watchdog, the Human Fertilisation and Embryology Authority, originally ruled that saviour siblings could be tissue-typed only if they were also being tested for a disease. It later reversed its decision, with the emergence of new evidence that the technique appears safe.

> **'We are all on the slippery slope. The question we should be asking is: skis or crampons?'**
>
> **John Harris, Professor of bioethics, University of Manchester**

A slippery slope? A further objection to PGD rests on a different argument: that while it is easy to sympathize with families who want to avoid having a baby with a serious disease, or to find a donor for a much-loved sick child, this sets an alarming precedent. Allowing even these applications, it is said, puts society on a slippery slope that will lead to embryos being screened for genes that affect intelligence, height or good looks. Children will come to be seen as commodities – at least, by those who can afford the technology.

It is perfectly possible, however, for societies to decide that PGD is permissible for some purposes, but not for others. The UK, indeed, has banned its use to choose a baby's sex for social reasons, or for deliberate selection of disabled children, but not for preventing disease.

The technology's dystopian potential is also firmly limited by science. First, PGD always requires IVF, which holds little appeal for the naturally fertile. Then there is the issue of what to look for. The sort of traits that pushy parents might want – such as intellectual or athletic ability – are governed by dozens of genes that interact in complex ways, and by environmental factors as well. It is impossibly difficult to screen for them all, or to guarantee the desired outcome.

There is also the matter of the raw material. Embryologists using PGD can work only with what nature provides, which means genes already carried by each parent. It is all very well ordering a designer baby with the brains of Stephen Hawking and the body of Kate Moss. If mum and dad don't have those qualities, it isn't going to happen.

PGD is a great tool for preventing genetic diseases that cascade down the generations, blighting whole families with misery and suffering. But it is entirely unsuitable for mass production of babies-to-order.

the condensed idea
Screening is not designing

42 Brave new worlds

Francis Fukuyama: 'What is ultimately at stake with biotechnology is . . . the very grounding of the human moral sense.'

In 1932, Aldous Huxley published a book that was to become a template for dystopian ideas about the future. In *Brave New World*, society was divided into five castes, from the dominant Alphas to the subservient Epsilons. Each individual was raised in an artificial womb in a 'hatchery', and then indoctrinated to accept his or her place in society. The lower orders were kept content with promiscuous sex and a hallucinogenic drug called soma. Comfort and order had abolished ambition and art, love and family, individuality and intellectual curiosity, even free will.

It was not a vision founded on the potential evils of genetics. Huxley was writing two decades before the double helix, and emphasized the horrors of extreme social conditioning rather than eugenics. As the science writer Matt Ridley has highlighted, his was 'an environmental not a genetic hell'. Even so, its themes have been taken up again and again by those who fear how genetic advances might threaten valued aspects of the human condition. Cloning, genetic engineering and DNA testing are often claimed to advance us towards a brave new world, in which liberty of body and mind could be lost.

Perhaps the best-known example is the 1997 movie *Gattaca*, the title of which is drawn from the four letters of the genetic code. The privileged classes use embryo screening to have 'valid' children of the best possible genetic quality, who monopolize society at the expense of the 'in-valids' –

timeline

1932
Aldous Huxley (1894–1963) publishes *Brave New World*

1997
Release of movie *Gattaca*

a genetic underclass. Kazuo Ishiguro has explored a slightly different theme in his novel *Never Let Me Go* (2005), in which the organs of cloned children are harvested as needed to prolong the lives of the people from whom they were created.

Our post-human future The notion that biotechnology threatens human values is not confined to fiction. The argument is also made by philosophers who want to limit the uses to which genetics is put. From the right, Francis Fukuyama of Johns Hopkins University has coined the idea of a 'post-human future', in which tinkering with DNA could upset delicate moral and ethical systems that are founded on an evolved, universal human nature.

Even well-intentioned applications of genetic technology to treat and prevent disease or suffering, he says, could undermine the idea that we are created equal, a founding principle of liberal democracy. His arguments are echoed by conservative bioethicists such as Leon Kass, who view cloning and germline genetic engineering as an assault on the dignity that separates humans from other animals.

Many of these views are shared by figures of the left, such as the philosopher Jürgen Habermas and the environmentalist Jeremy Rifkin, who fear that biotechnology threatens the 'species ethic' that makes us respect the lives, intentions and aspirations of other humans. Bill McKibben, in his 2003 book *Enough: Staying Human in an Engineered Age*, makes a related point, that enhancement technologies break the link between people and their past, calling into question what it means to be human. He is particularly critical of germline genetic engineering, which he believes will make children question whether their achievements and aspirations are their own, or the result of genetic impulses implanted by their parents.

A common concern is that genetic technologies will be disproportionately available to the wealthy, creating a *Gattaca*-style DNA divide. The rich will be free to improve their genomes and those of their children, to prolong their lives and entrench social advantage. The poor will be left

2001
First drafts of human genome completed

2002
Francis Fukuyama publishes *Our Posthuman Future*

2005
Kazuo Ishiguro publishes *Never Let Me Go*

Immortality?

Some transhumanists, such as the British theorist Aubrey de Grey, believe biotechnologies could eventually end ageing. Stem cells and genetic manipulation could allow us to replace body parts as they wear out. Even death, de Grey thinks, is an engineering challenge waiting to be overcome.

Most mainstream biologists are sceptical about this, chiefly because the abolition of ageing would have to buck natural selection. Once we pass the age at which we no longer reproduce, the evolutionary pressures that encourage good health no longer apply. The sort of genetic errors that contribute to cancer and heart disease in our later years have not been weeded from the gene pool, because their deleterious effects start only once they have already been passed on. We are not designed to live forever.

Extreme longevity could also have unpleasant social consequences, of which overpopulation is only the most obvious. As Richard Dawkins has noted, it would also transform attitudes to risk. Even if we can stop ourselves dying of old age and disease, we will still be vulnerable to accidents. With a life expectancy of 80, it makes sense to take the odd chance. With a life expectancy of 800, even crossing the road might look unacceptably dangerous.

behind, creating the conditions for bitter conflict between genetic haves and have-nots. Many disabled people also feel that such technology marginalizes them as second-class citizens who should not exist.

Transhumanism Supporters of human biotechnology tend to counter these arguments with three questions. Why not? Are such worries really justified? And can progress anyway be stopped?

On the first question, figures such as the philosophers John Harris and Julian Savulescu, and the authors Ronald Bailey and Gregory Stock, take a libertarian position. If stem cell therapies, screening techniques and even genetic engineering are sufficiently safe, and do no harm to others, there is no compelling reason why people should not be free to use them. Most people happily embrace medicines that improve the length and quality of their lives, and those of their families, and techniques that involve DNA

‘Given how much people value life, to protect them from premature death or give them a longer healthier life expectancy seems to me a powerful obligation.’

John Harris

or reproduction should not be special cases. The choice of whether or not to use them is best left to individuals, not to society.

To the second question, many biologists and ethicists would answer no, but for two very different reasons. One group, sometimes known as transhumanists, argue that genetic technologies are not to be feared but to be celebrated. If science can help people to suffer less and achieve more, isn't that a positive thing? Harris has gone so far as to suggest that it is not only morally justified to seek better ways of fighting sickness and disability and improving human bodies and minds, but morally obligatory.

Others point out that many concerns rest on a misunderstanding of genetics, which places too much weight on its power to determine. DNA is of course important to human nature, but it does not prescribe it as it does the amino acid sequence of insulin. As we saw in Chapter 17, both genes and the environment matter to the human condition. It is impossible to boil down individual identities, or that of our species, to this gene or that. As Kenan Malik, a British science writer, wrote in review of a Fukuyama book, humanity's uniqueness lies in our ability to be conscious agents. It is not likely to be possible to engineer that away.

To the final question, transhumanists point to the lessons of history. Once technologies have been invented, only rarely have they been abandoned, and never for very long. If a genetic technique holds out the hope of a better existence, whether that comes through treating a disease or enhancing an ability, people will always want to take advantage of it, and some eventually will. It might be better to manage these aspirations through regulation than to pursue unworkable bans. The real challenge is to ensure fair and safe access to exciting technologies, not to find ways of holding them back.

the condensed idea
Genetics is both opportunity and threat

43 Genes and insurance

Søren Holm: 'If we accept that life or health insurers can legitimately seek and obtain other kinds of health information that predicts insurance risk, then we should also accept that they can seek genetic information that is predictive in the same way. There is no reason for treating genetic information differently.'

When Aids first emerged in the United States, many gay men were initially sceptical about the value of taking a test. In the absence of effective treatment, some simply preferred not to know their HIV status, but others had a more practical objection. A positive result would not only be stigmatizing, it would make it virtually impossible to buy insurance.

The advance of genetic healthcare is today being delayed by similar fears. DNA tests that assess individuals' risk of developing particular diseases have great potential in preventive medicine, but the existence of such data can be threatening, too. In the hands of an insurance company, it could also be used to deny access to health cover, or to the life policies that are required when obtaining a mortgage.

The genetic threat Commercial insurance works on the basis of pooled risk. By paying premiums, customers together build a fund that will compensate those who are unlucky enough to die young or fall ill. Some contributors will collect on their policies, while others never will. Insurers want to sign up as many likely non-collectors as possible, while keeping high-risk clients to a minimum. They therefore employ actuaries to assess

timeline

1993

Isolation of Huntington's disease mutation

individuals' risk before setting premiums, on the basis of information such as smoking and sex, postcode and profession.

This system works, in part, because neither party can be entirely confident about the future. Add genetics to the mix, and it could become dangerously unbalanced. If insurers can discover the results of DNA tests, they could use these to charge elevated premiums, or to refuse cover altogether, to those with high-risk genomes.

To figures such as John Sulston, a pioneer of the Human Genome Project, this would be unfair and unethical. At present, actuarial decisions are mostly based on factors over which individuals have some control, such as smoking or address. Nobody, however, has any influence over the genes they inherit from their parents. To allow insurers access to these could render some people uninsurable through no fault of their own. It would also encourage people to avoid genetic tests that might be beneficial to their health, as happened with HIV, or to refuse to participate in genetic research.

Genetic information, too, is rarely deterministic. Most medical conditions are not like Huntington's, in which a mutation leads invariably to disease and death. The contribution of genes to disease is often incompletely understood. The leads that genetic tests provide will often be inaccurate, adding to the injustice of enforced disclosure.

Such arguments have resonated with public opinion. In the UK, the insurance industry has agreed a voluntary moratorium on demanding genetic data, with the solitary exception of the highly reliable Huntington's test. In May 2008, President Bush signed the Genetic Information Non-discrimination Act (GINA), which bans the use of DNA tests by employers or health insurers.

Double-edged injustice Genetic information, however, can lead to injustice in the other direction, too. If customers acquire important data about their probable future health, yet have no obligation to declare it, they can play the system by buying policies they know they will probably cash in. A study at Duke University in the US has shown that people who

2001
First drafts of human genome completed
UK insurers agree moratorium on use of genetic tests

2008
US Congress passes Genetic Non-Discrimination Act

discover they have a genetic variant that raises Alzheimer's risk become more likely to buy insurance to cover long-term nursing care. This is not only unfair to companies but also discriminates against other policyholders, who must pay higher premiums as a result.

These issues have led some commentators, such as the philosopher Martin O'Neill, to suggest that voluntary mutual insurance may not survive in its present form. As it is unjust to force individuals to disclose genetic test results, or to deny them to companies, the state may have to step in. A compulsory system in which everybody contributes, independently of individual risk, could be required to ensure equality of access. This, indeed, is already the model used by socialized health systems such as Britain's NHS.

Is genetic information really special? The challenge, however, may be less serious than it initially appears. First, the principle that genetic information can be considered by insurers is already well established. Car, life, health and accident policies all discriminate on the basis of a gene called SRY – the one that makes you male. Men can no more choose whether to have it than a woman can choose to inherit a BRCA mutation. Yet sex is an accepted tool with which actuaries set premiums, and few people think this is wrong.

Genetic privacy

Insurance companies are not the only institutions that might be interested in genomic information. Some businesses will wish to screen potential employees for general good health, for genetic aptitude (though Chapter 22 demonstrates why this could be misleading). Some air forces already test potential pilots for the sickle-cell anaemia mutation, as carrying even a single copy raises the risk of a blackout. Police keep specialized genetic databases, and it is easy to envisage circumstances in which they would also seek access to medical registers to trace suspects. Even private individuals could find others' genetic information useful, to confirm paternity or draw up family trees.

As personal sequencing becomes more affordable, and more and more people's genomes are mapped for medical reasons, these privacy issues will become telling. Society is going to have to think hard about whether and how this information should be kept confidential, and about who has a right to access.

Genetic conditions that develop early in life, such as haemophilia or muscular dystrophy, are also usually considered fair game. People would be expected to declare such a diagnosis, so why not a mutation that will cause Huntington's later? Neither is existing genetic discrimination confined to disorders that invariably cause disease, or to genetic tests. As Søren Holm, of the University of Cardiff, points out, insurers are quite entitled to ask about family history – a proxy for genes – and often raise premiums for people who have lost relatives to conditions such as early-onset heart disease.

If anything, genetic discrimination of this sort is less accurate than that based on the outcome of DNA tests. People who know they have a parent with Huntington's already struggle to get insurance because they have a 50 per cent chance of having inherited the mutation. Each individual's true risk, however, is actually 100 per cent or zero: they either have the mutation or they don't. A test can at least decide this, making some people insurable while changing nothing for the less fortunate.

The problems posed by genetic testing may well apply only while our knowledge of genetic influences on disease remains preliminary and incomplete. At the moment, it would probably be unfair for companies to assess the few variants that are known to affect risk slightly, as so many others are involved. Yet when a fuller picture is known, many of the problems are likely to fall away.

Almost everybody will have a profile that predisposes a little to some conditions and protects against others. Insurers have to insure people if they want to make money, yet nobody has a perfect genome. They will have to take on customers with known genetic risks, at reasonable premiums, or go bust. Governments may well want to make special arrangements for unlucky individuals with rare major mutations, whom the industry will not want to touch. A little genetic discrimination, though, need not break the whole system.

‘Americans can finally take advantage of the tremendous potential of genetic research without the fear that their own genetic information will be used against them.’

Louise Slaughter, US Representative, on the GINA

the condensed idea
Insurance can survive genetics

44 Gene patenting

John Sulston: 'A genome sequence is a clear-cut case of public domain material.'

In 2001, a biotechnology company named Myriad Genetics was granted European patent number EP699754. It covered the DNA sequence of the BRCA1 gene, and a test for mutations that can raise a woman's lifetime risk of developing breast cancer to 80 per cent. It has become a symbol of one of the most vexed questions in biology: how intellectual property laws should be applied to genetics.

When the BRCA1 patent was awarded, the decision outraged public-sector scientists. A charity-funded team had laid much of the groundwork for the isolation of the gene. Bruce Ponder, its leader, has claimed to have been '100 yards from the finishing post' when Myriad noted this progress, raised millions on the capital markets to complete the sequence, and applied for a patent before its rivals could publish.

Though the public group released the gene's code soon afterwards, Myriad's patent had priority. It thus acquired a monopoly over BRCA1 testing (though an attempt to claim the BRCA2 gene too failed, at least in Europe), and the right to charge what it liked for a medical service that can save women's lives.

How patents work The patent system exists so that inventors can protect their ability to profit from their work, without having to keep its details secret. In return for publishing the specifications of an innovation, patent-holders are granted exclusive rights over commercial applications, usually for 20 years. To qualify for patent protection, an advance must generally satisfy three criteria. It must be new, it must involve an inventive step that is not obvious, and it must be suitable for commercialization.

1993
James Watson resigns from Human Genome Project in row over gene patenting

1995
Discovery of BRCA1 mutation

The system is fundamental to innovation, providing a powerful incentive for companies to invest in research and development, and then to share their discoveries. Few people dispute that individuals, institutions and businesses should be able to prevent their true inventions from being unfairly exploited by others. Drugs, medical procedures and diagnostic tests are all clearly patentable, and most scientists accept that this rewards achievement and stimulates research. But where genes, proteins and cells are concerned, the subject becomes much more contentious.

Organisms and cells

While organisms that exist in nature cannot be patented, the situation for genetically modified life is less clear. Most jurisdictions will award patents on GM plants, such as Bt cotton (see Chapter 32), and on products of GM bacteria, such as recombinant insulin. GM animals are more controversial, as many lawyers question whether higher organisms can be considered intellectual property. Europe and Canada have granted a patent covering the OncoMouse (see Chapter 33), which is widely used in cancer research, but with tight restrictions on its terms.

Patents on tissue, such as embryonic stem cells, are similarly controversial. Stem cells themselves cannot be patented as they occur naturally, but methods of extracting them can be. The standard technique, developed by Jamie Thomson of the University of Wisconsin, has been awarded a patent, but this has been challenged on the grounds that it was obvious. It was revoked, then partially reinstated, and appeals are ongoing.

There is a consensus that naturally occurring organisms cannot be patented, as they must be discovered, not invented. But what about their components? Supporters of genetic patenting argue that finding genes is not a trivial process – until recently, it took years of research. Patent protection is said to encourage investment in such study, and thus to encourage the development of genetic medicine.

2001
Myriad Genetics granted BRCA1 patent

2006
More than 4,000 human genes patented

2007
Myriad BRCA1 patent struck out in Europe

Critics such as Nobel prizewinner John Sulston, however, take a different line – that the genomes of all plants and animals, and particularly that of *Homo sapiens*, are entities that existed long before it became possible to read them. While sequencing techniques are new and inventive, genes themselves are not. It should thus be possible to patent genetic technologies, such as methods for mapping genes and probes to test for their presence, but not the genes themselves. These should remain the common property of humankind. In 1993, James Watson resigned as leader of the Human Genome Project after a fierce row with Bernadine Healy, then the director of the US National Institutes of Health, over her plans to file genetic patents.

Sulston, Watson and other opponents of gene patenting think excessive intellectual property protection insidious, because it impedes valuable research. If it is always necessary to buy licences to study particular genes, fewer groups are going to do so. Too broad an interpretation of patent rights would also drive up the price of genetic products, such as the BRCA1 test over which Myriad won control, denying access to many patients.

What is more, as publicly- and charitably-funded scientists tend to publish their sequencing work as they go along, less scrupulous companies can use these free data to accelerate gene-discovery programmes, and then patent the results. That is just what Myriad was accused of doing, and Craig Venter has noted with irony that the Human Genome Project's daily data-release scheme helped businesses to file hundreds of patents.

The land grab held back When gene sequencing came of age in the 1990s, it triggered a genetic gold rush, as scores of companies and institutions moved swiftly to patent swathes of human DNA – and thousands of these applications were granted. In 2006, a paper published in *Science* estimated that more than 4,000

Patients' patent rights

Genetic discoveries require raw material, and many people feel that those who donate DNA and tissue for use in medical research should share in its profits. Such donors, however, have very few rights in law. In the 1970s, a University of California Los Angeles clinic treated a leukaemia patient called John Moore, and used his tissue to create a cell line for cancer research. It patented this tissue in 1981, and Moore sued for a share. In 1990, the Supreme Court of California ruled against him, on the grounds that cells removed from his body with his consent were no longer his property.

‘If patents are granted that are too broad in scope, they block other researchers from carrying out related work and so hold up the development of medicines. This is tremendously bad for science, but the ultimate losers are the patients.’

John Enderby, Royal Society

human genes – almost a fifth of the known total – had been patented. Many of these patents are held by public or charitable institutions, filed in part so that private companies could not control them – that is what Healy had hoped to achieve by patenting the results of NIH work. But almost two-thirds are held privately, and one company, Incyte, has rights over about 2,000.

The situation, however, is starting to change. In 2000, President Clinton declared that the human genome itself was unpatentable, a statement that depressed the prices of biotech shares. The tide against broad genetic patents then turned further with reports from respected scientific organizations such as the Nuffield Council on Bioethics and the Royal Society, which argued that genes are not novel, and that speculative patents without commercial applications were holding back medical research. The outcome of the race between public and private consortiums to sequence the human genome, and Celera's ultimately failed attempt to restrict access to its data, also reinforced the idea that genes are common property."

Many patents on genes are now collapsing under legal challenges. One of them has been Myriad's over BRCA1. In 2004, the European Patent Office ruled that the company's final application had not been novel, as it included data that had already been published by the charitable team. It struck out the patent, and in 2007 rejected Myriad's appeal. The cost of BRCA1 tests is now coming down. Such cases are leading many biotech companies away from patenting genes, or aggressively enforcing the ones they have. An intellectual property system is emerging that applies to genetic technology, but not to genes themselves.

the condensed idea
Genes are not inventions

45 Junk DNA

Manolis Dermitzakis, ENCODE consortium: 'If you think of the letters that make up the human genome as the alphabet, then you can think of genes as the verbs. We're identifying all of the other grammatical elements and the syntax of the language we need to read the genetic code completely.'

The human genome contains three billion base pairs, the DNA letters in which the code of life is written. Yet only a tiny proportion of these letters – no more than 2 per cent – are actually used to write our 21,500 or so genes. The remainder, which makes none of the proteins that drive the chemical reactions of life, has long been something of a mystery. Its apparent lack of function has led it to be dubbed 'junk DNA'.

The presence of large tracts of DNA with no purpose, however, would present an evolutionary puzzle. It takes energy to copy DNA, and if the vast quantities of junk found in all organisms were truly useless, it ought not to have survived the attentions of natural selection. Individuals that successfully eliminated inert genetic material should have an advantage over those that do not, producing thrifty genomes of more manageable size. The fact that they have not done so suggests that junk DNA is somehow important.

A further clue to its significance was revealed when the Human Genome Project found many fewer protein-coding genes than the 100,000 that had once been predicted. This tally seemed far too low to explain all the differences between humans and other organisms, indicating that the genome must amount to more than the sum of its genes. Beyond the genes, what remained was the junk, which geneticists have now started to observe with fresh eyes.

timeline

1941
Discovery that genes make proteins

1953
Identification of structure of DNA

What's in the junk? Much of our junk DNA has origins that have been relatively simple to establish. A very large part of it belonged originally to viruses, which have incorporated their own genetic codes into our genome in order to reproduce. These human endogenous retroviruses are currently thought to make up about 8 per cent of the total: they account for more of the book of humanity than human genes.

The legacy of our viral ancestors can also be seen in so-called retrotransposons. These repetitive chunks of DNA, which were originally deposited by viruses, have the ability to copy themselves into the human genome again and again, using an enzyme called reverse transcriptase. The commonest class are known as LINEs (for long interspersed nuclear elements) and on current estimates, they account for around 21 per cent of all human DNA. Shorter retrotransposons, the commonest of which is the Alu family, make up even more of the genome, and still smaller elements include the short tandem repeats, which are used in DNA fingerprinting.

Other types of non-coding DNA include the introns that separate the protein-coding sections of genes, and the centromeres and telomeres that occur at the middle and ends of chromosomes, respectively. There are also pseuodogenes – the rusting wrecks of genes that were important in our ancestors, but which have decayed through mutation. Hundreds of these genetic fossils can be found in the human genome (see box).

What does it do? In some ways, the continued presence of this junk DNA is not surprising: DNA is 'selfish', and will replicate itself regardless of utility to its host organism. But for it to withstand natural selection, some of it must surely be functional. Further evidence for a biological role comes from more than 500 regions of junk DNA that are highly conserved from species to species. These have probably been preserved because they perform a vital task, to which mutations are catastrophic.

One hypothesis for junk DNA's role is that it protects genes. If the genome contained nothing but protein-coding elements, many of these would be broken up and rendered meaningless by recombination errors. Non-coding DNA could provide a buffer, reducing the probability that a critical gene

1961
Discovery of triplet genetic code

1984
Development of genetic fingerprinting

2001
First drafts of human genome reveal surprisingly few genes

2007
ENCODE consortium shows 9 per cent of genome is transcribed

Fossil genes

Some of our junk DNA is made up of 'pseudogenes' – sequences that were once working genes, but which have lost their ability to make proteins though lack of use. They are genetic fossils, which tell the story of evolution as faithfully as fossilized bones.

When important genes acquire mutations, these are usually weeded out by natural selection because they put individuals that carry them at a disadvantage. But when a gene codes for a protein that a species no longer needs, that disadvantage no longer applies. Animals that live underground, such as moles, will not suffer if a mutation knocks out an eyesight gene. As mutation occurs randomly, but at a consistent rate, such luxury genes will inevitably decay over time. Defunct versions of these genes, though, will continue to be preserved in their genome.

A good human example is the Vr1 family of genes, which are involved in detecting scent. Mice have more than 160 functional Vr1 genes, while people have just five. The dead Vr1 genes have not vanished from the human genome – they have fossilized, providing proof of our shared evolutionary heritage with mice.

will be damaged. Another idea, also founded on recombination, is that it provides a reservoir from which new genes can evolve. When chromosomes cross over, some bits of junk might fall together in useful new combinations. This would make the junk analogy rather apposite – junk, after all, is not thrown away, but stored in case we can put it to good use in the future.

It is now clear, however, that much of our junk DNA is misnamed – it is not surplus to requirements at all, but carries out specialized and significant tasks. Large parts of it are thought to be involved in regulating gene activity, in generating messages that tell coding parts of the genome when and how to operate, and when to keep quiet.

The most telling evidence for this hidden biological function has emerged from the ENCODE (Encyclopedia of DNA Elements) consortium. This international effort to study the workings of the whole genome, not just

‘It’s no longer the neat and tidy genome we thought we had. It would now take a very brave person to call non-coding DNA junk’

John Greally, Albert Einstein College of Medicine

the genes, is effectively compiling a ‘parts list’ of the DNA that is biologically active in the body. Its pilot phase, which reported in 2007, has looked in detail at 30 million base pairs – 1 per cent of the total.

What it found was remarkable. While only about 2 per cent of the genome is made up of genes, at least 9 per cent of it is transcribed into RNA – a sign that much of it is biologically active. Only a small proportion of this transcribed RNA is the messenger RNA that carries protein-making instructions. Junk DNA generates different kinds of RNA, the range of which will be explored in Chapter 48. These molecules, in turn, modify the expression of genes and proteins, to fine-tune human metabolism.

This fine-tuning has a profound effect on physiology. Single-letter DNA changes that influence the risk of disease have been found in non-coding parts of the genome, as well as genes. A rare mutation in a gene called MC4R, for instance, causes childhood obesity, but people with normal versions are also more susceptible to weight gain if they inherit a common variant in the surrounding junk. The variant seems to lie in a region that regulates MC4R, altering its activity.

Variation in non-coding DNA could also explain differences between species. Some 99 per cent of human and chimpanzee genes are identical, compared with only 96 per cent of all DNA. As there is so much more diversity in the junk, this may well underlie uniquely human traits such as intelligence and language. The notion that protein-coding genes are the only contents of the genome that matter is manifestly mistaken.

the condensed idea
Junk isn’t rubbish

46 Copy number variation

Matthew Hurles: 'Each one of us has a unique pattern of gains and losses of complete sections of DNA. We now appreciate the immense contribution of this phenomenon to genetic differences between individuals.'

It has become a commonplace that humans are 99.9 per cent identical at a genetic level. The mapping of the human genome revealed that while the human genome contains three billion DNA base pairs, only around three million of these, or 0.1 per cent, typically vary in spelling. These one-letter changes are the single nucleotide polymorphisms or SNPs. A little genetic variation seems to go a very long way.

This estimate of genetic difference, however, has since been revealed to be wrong. SNPs, it turns out, are not the only way in which genomes vary. Whole genes and fragments of genes can also be duplicated, deleted, reversed and inserted into the genome. This new type of variation was found in 2006 to be extremely common, and it has as great a bearing on biology and health as the conventional kind.

This copy number variation, also known as structural variation, suggests that the average genetic difference between any two people is not 0.1 per cent, the figure derived from SNPs. It is actually at least three times greater, at 0.3 per cent or more. This may partially explain why so few SNPs can produce such extensive human diversity: our knowledge of the genome's variability was incomplete. It is an insight that has forced a reappraisal of how DNA makes every one of us – and our species – unique.

1941
Discovery that genes make proteins

1953
Identification of structure of DNA

Duplications and deletions The standard model of genetics is that everybody inherits two copies of the genetic sequence, one from each parent. Yet a research team led by Matthew Hurles, of the Wellcome Trust Sanger Institute, and Charles Lee, of Harvard Medical School, has established that this is too simple. When they started to look in detail at the genomes of 270 people, originally recruited for the HapMap project we met in Chapter 19, they found that the double-copy paradigm was not universal at all.

Across about 12 per cent of the genome, large portions of DNA, ranging in size from 10,000 base pairs to 5 million, were sometimes repeated or entirely absent. While most people have just two copies of these sequences, some have one or none, while others have several – as many as five to ten copies in some cases. Strings of DNA can also be inserted out of place, or inverted so they read back to front. The genome varies considerably in structure, as well as in spelling.

Investigating copy number variation

The first wave of genome-wide association studies, the powerful new tools for finding genes that affect disease that were introduced in Chapter 19, looked only at SNPs. The growing understanding that copy number variation is at least as important is changing the way this research is being designed. In April 2008, the Wellcome Trust announced a £30 million grant to fund a second phase of its successful Case Control Consortium, which will investigate two dozen new diseases. This time, it will cast the net beyond SNPs, using gene chips that can detect copy number variants as well.

1961 Discovery of triplet genetic code

2001 First drafts of human genome

2006 Discovery of large-scale copy number variation

It has long been known that some stretches of human DNA are occasionally duplicated or deleted, as with the extra copy of chromosome 21 that causes Down's syndrome. Such changes, however, had been thought to be rare, and serious in their consequences. It is now known that variations of this sort are common – more common, in fact, than conventional SNPs.

Sometimes, this structural variation is trivial – as with SNPs, certain changes make no difference to genetic function. But it can be strongly linked to altered physiology or susceptibility to disease, and may also account for differences between species. Once copy number variation is brought into play, we share just 96 to 97 per cent of our DNA with chimps, not the 99 per cent estimated from spelling alone.

Copy number and disease The most exciting implications of copy number variation lie in its consequences for disease. Now that scientists have become aware that it is something worth looking at, all manner of associations are starting to emerge between individual health and DNA deletions, duplications, insertions and inversions.

A gene called CCL3L1, of which some Africans have multiple copies, is one of the most interesting early manifestations of this phenomenon. Those with a high copy number, it appears, are less susceptible to infection with HIV. While it is not yet known precisely how or why this happens, a working hypothesis is that extra copies boost production of a protein that is important to HIV resistance. This promises to open new approaches to treating the virus, and to preventing its spread.

Other copy number variations that have been reliably linked to disease include genes known as FCGR3B, in which a low number predisposes to the auto-immune disorder lupus, and EGFR, which is often

New genetic territory

The Human Genome Project did not complete a comprehensive map of our genetic code. It produced an average sequence that provides a reference point against which scientists can compare the DNA of individuals and other species. Studies of copy number variation are now revealing whole segments of DNA that do not appear in this reference genome, but which are nevertheless reasonably common. A 2008 study that looked in detail at eight people's genomes found no fewer than 525 new sequences that are sometimes inserted into the code. Many more probably remain undiscovered.

repeated in patients with non-small cell lung cancer. People of south east Asian origin often tend to have multiple copies of another gene, which seems to offer a degree of protection against malaria. An examination of structural variation in genes expressed in the brain has found possible associations with 17 conditions of the nervous system, including Parkinson's and Alzheimer's.

Copy number variation is also offering insights into the genetic origins of two of the most confusing conditions in which inheritance is involved: schizophrenia and autism. Twin and family studies have proved that both disorders are highly heritable, yet the search for genetic variants and mutations that are responsible has drawn little success. Recent research, much of it led by Jonathan Sebat at Cold Spring Harbor Laboratory, has suggested that copy number variation is often involved – particularly in sporadic cases that occur among people with no family history of these conditions.

'Copy number variations are the most common cause of autism we can identify today, by far.'

Arthur Beaudet, Baylor College of Medicine

Deletions or duplications in certain 'hot spots' of the genome are much more common among children with autistic spectrum disorders than in the general population. Many have variations that are absent in their non-autistic parents. For schizophrenia, Sebat's group found that rare copy number variations are present in 15 per cent of people who developed the mental illness as adults, and among 20 per cent of teenage patients, compared with just 5 per cent of healthy controls. Many of the changes in copy number that affect both disorders may be unique to the individuals who carry them – which could explain why their genetic roots have been so hard to pin down.

These discoveries are changing the way in which scientists think about genetic diversity. As Hurles puts it: 'The variation that researchers had seen before was simply the tip of the iceberg, while the bulk lay submerged, undetected.' This vast repository of difference is only beginning to give up its secrets, but science now knows that it is there.

the condensed idea
Genes vary in structure as well as spelling

47 Epigenetics

Marcus Pembrey: 'We are changing the view of what inheritance is. You can't, in life, in ordinary development and living, separate out the gene from the environmental effect. They're so intertwined.'

In the autumn of 1944, railway workers in the Netherlands, then under German occupation, went on strike to assist the advancing Allies. When the initial British and American assault failed, the Nazis retaliated by imposing a devastating food embargo. At least 20,000 Dutch citizens starved or died of malnutrition in the ensuing famine.

The effects of the *Hongerwinter* or 'Hunger Winter' were to last long beyond the country's liberation in 1945. Mothers who were pregnant during the famine had children with an elevated risk of a wide range of health problems, including diabetes, obesity and cardiovascular disease. In some cases, even their grandchildren were more likely to be born underweight. While damage to the first generation's health might be explained by malnutrition during pregnancy, the Netherlands was a rich nation by the time the second generation was born. Yet an inherited effect remained.

The story of the Dutch famine is not unique. The village of Överkalix, in northern Sweden, boasts meticulous historical records of harvests, births and deaths, which have allowed Marcus Pembrey, of the Institute of Child Health in London, to conduct an exhaustive study of food availability and life expectancy. He found that when boys grew up during periods of plenty, their grandsons were more likely to die at a young age. Further analysis revealed that this reflected a predisposition to diabetes and heart disease, and confirmed that the effect was passed only through the male line.

1802

Lamarck proposes inheritance of acquired characteristics

1990s

Identification of epigenetic effects in mice

Both examples suggest that people's health can be affected by the diets followed by their grandparents. Yet according to orthodox evolutionary theory, such an effect should be impossible. Acquired characteristics are not supposed to be inherited – that was the Lamarckian heresy, which has not been fashionable since Darwin.

Genetic memory The Dutch and Swedish experiences can be explained by a phenomenon known as epigenetics, by which the genome appears to 'remember' certain environmental influences to which it has been exposed. Normally, these epigenetic effects act only on the somatic cells of the adult body, switching genes off or otherwise adjusting their activity. Some, however, can also alter sperm and eggs, to be inherited by future generations. Acquired characteristics, it turns out, can sometimes be passed on after all.

Suicide

Epigenetic effects may explain how dreadful experiences leave their mark on human behaviour, making adults more likely to suffer depression and even to commit suicide. A team led by Moshe Szyf, of McGill University, examined DNA from the brains of 13 men who committed suicide, and found that while their genetic sequences were normal, their epigenetic programming was different from that seen in men who had died of other causes. All the men in the study had been abused as children, which could have triggered this epigenetic change. 'It's possible the changes in epigenetic markers were caused by the exposure to childhood abuse,' Professor Szyf says.

Epigenetics owes its prefix to the ancient Greek for 'over' or 'on', and it generally relies on two broad mechanisms. One is methylation, the process we met in Chapter 29, which silences genes by adding part of a molecule

2002
Human epigenetic inheritance advanced as explanation for Swedish life expectancy

2004
Approval of first epigenetic drug

2008
Identification of epigenetic markers in brains of suicide victims

called a methyl group to the DNA base cytosine, or C. The other is modification of chromatin, the combination of DNA and histones (kinds of protein), of which chromosomes are made up. Changes to chromatin structure can affect which genes are made available for transcription into messenger RNA and protein, and which are hidden away out of reach. In neither case is the actual sequence of DNA altered at all, but changes in its organization can still be passed on from one cell to its offspring.

These epigenetic processes are central to normal growth, development and metabolism. Every cell contains the full set of genetic instructions that are needed by every type of tissue, and epigenetics determines which of these are actually issued and executed. It ensures that genes that are required for rapid cell division in the embryo are switched off in the adult body, so they cannot cause cancer, and controls the patterns of gene expression that tell a cell it belongs to a kidney or brain.

Epigenetic effects also allow nurture to guide nature, by changing the way genes act in the body in response to environmental cues. Experiments with mice have demonstrated this lucidly: changes in diet while females are pregnant affect the coat colour of their offspring, by modifying the way genes are methylated. This effect, indeed, could account for the surprising observation that many cloned animals differ from their parents in colour and markings. While their genomes are identical, their 'epigenomes' are not.

Normally, such epigenetic changes are stripped away from the genome during embryonic development, so that they are not passed on to offspring. But they can sometimes be retained, causing environmental effects on health and behaviour that cascade down the generations. That could account for what happened in Holland and Sweden. Parental diets seem to have changed the epigenetic programming of their children and grandchildren, to modify metabolism to cope with the prevailing nutritional environment. This, in turn, has influenced health risks such as diabetes.

The importance of the epigenome As with copy number variation and junk DNA, science is starting to understand that epigenetic effects are just as significant to biology as conventional genetic mutations. Epigenetics, for example, plays an important role in cancer. Several chemicals are known to be carcinogenic, even though they are not mutagens that directly damage DNA. They induce epigenetic effects, silencing important tumour suppressor genes, or altering chromatin structure so that oncogenes become more active.

Epigenetic markings also ensure that when cancer cells divide, their daughter cells are cancerous too. An understanding of how these processes work could open a new approach to medicine. The first cancer drug that works by removing methylation, Vidaza, was approved by the US Food and Drug Administration in 2004.

The Human Epigenome Project, recently started by a European consortium, should help to bring more epigenetic therapies into medicine. This ambitious initiative aims to map all the possible methylation patterns of every gene in every type of human tissue. A pilot project has already achieved this for the major histo-compatibility complex – a cluster of genes on chromosome 6 that affect immune response.

Once these methylation sites have been identified, it should be possible to link variations to diseases, in much the same way as can be done for SNPs. Doctors, indeed, may well find their patients' epigenomes more useful than their genomes. The genetic code, as the early history of gene therapy has proved, is painfully difficult to correct in living organisms; methylation should be comparatively simple to undo. Many of the medicines of the future could be designed to exploit this natural method of genetic control to treat and prevent disease.

Stem cells

Though embryonic stem cells can grow into any type of tissue, their genetic code is no different from that of the specialized adult cells to which they give rise. Their uniquely malleable properties seem to be derived from their epigenetic character. Adult cells from the skin or bone carry all the genetic instructions that are needed to make any other cell type, but most of these instructions are switched off by epigenetics. Only in embryonic stem cells are all the genes required for pluripotency unmethylated and active.

It has recently become possible to reprogram adult cells into a pluripotent state (see Chapter 35), but only by replacing silenced genes with active copies – a technique which carries a risk of causing cancer. This hazard might plausibly be reduced by reprogramming these cells' epigenomes instead.

the condensed idea
The genome can remember

48 The RNA revolution

Chris Higgins, UK Medical Research Council: 'RNA interference provides a simple tool for manipulating gene expression in the laboratory, and has great promise for altering gene expression to treat diseases such as viral infections and cancer.'

Since its discovery by Friedrich Miescher, and especially since the discovery of its structure by Francis Crick and James Watson, DNA has been considered to be the queen of the nucleic acids. It is the stuff of which genes are made, the encoded language in which the instruction manual of life is written.

Ribonucleic acid suffers by comparison to this giant among molecules: RNA has often been seen as DNA's servant. It is the signalling chemical that runs cellular errands for its master, and the fetcher and carrier that collects amino acids so that DNA's muse can find expression in protein.

RNA, however, now looks rather more interesting than the first generation of molecular biologists had assumed. So interesting, in fact, that some scientists think it necessary to reassess the priority of the two nucleic acids that between them drive every form of life on the planet. DNA might contain the genome's core information, but it is through its chemical sister that it shapes organisms and their life cycles. RNA is anything but passive: it is a dynamic and versatile molecule that exists in dozens of forms, the vital functions of which science is only beginning to understand. It may even be the root of life itself.

timeline

1868
Friedrich Miescher discovers DNA and RNA

1960
Messenger RNA revealed as 'adaptor molecule'

The many faces of RNA We have already met the most basic type of RNA – the single-stranded messenger RNA (mRNA) molecules into which DNA is transcribed, and which carry instructions for making proteins. Only about 2 per cent of our RNA, however, is mRNA.

Many other varieties are involved in protein manufacture alone. The important bits of mRNA – the exons – are interspersed with stretches of nonsense called introns. An RNA-based structure called the spliceosome snips away the introns, and stitches the exons into a meaningful message. This then travels to a cell's ribosomes, or protein factories, which are themselves made chiefly of ribosomal RNA, another specialized type. Transfer RNA, a cross-shaped variety, then identifies and collects amino acids for threading onto protein chains.

RNA is not just a protein-making tool. It also comes in small molecules such as micro RNAs (miRNAs), which are tiny single strands of between 21 and 23 bases in length. These are transcribed from DNA, but from the junk DNA that does not code for proteins, and their function seems to be to regulate how genes work in the body. They switch them on and off, or tweak their activity so that levels of protein manufacture go up or down. These are now thought to explain much of the complexity of the human life.

The origin of life

The question of how life on Earth began, about four billion years ago, is one to which there remains no answer. One leading hypothesis is that some of the first self-replicating life forms, if not the first, were based on RNA. It is simpler than DNA, usually occurring in one strand rather than two, and it can both replicate itself and catalyse chemical reactions from surrounding molecules. This led figures such as American microbiologist Carl Woese and Francis Crick to suggest that primitive 'ribo-organisms' might have used chemicals from their environment to create new copies of themselves. Only later did life move beyond this 'RNA world', and start using the more robust DNA molecule to encode its genetic information.

1967
Carl Woese proposes RNA as basis of earliest forms of life

1990s
Discovery of RNA interference

2007
ENCODE consortium finds much more DNA is transcribed into RNA than was anticipated

There are thousands of different types of human miRNA – the total number may well exceed the tally of around 21,500 genes. Each can modify not only the activity of single genes, but that of groups of genes and other RNA molecules as well. This means that when thrown together in combination, miRNA can manipulate gene expression in subtle ways that are almost limitless. They allow a relatively small range of genes, many shared with other animals, plants and even microbes, to produce structures as complicated as the human brain. There is good evidence, indeed, that the number of miRNAs increases with an organism's degree of complexity. While people have only a couple of thousand more genes than do nematode worms, we have many times the number of miRNAs. These molecules seem to be responsible for building more sophisticated forms of life.

RNA interference Growing recognition of the importance of RNA is also shedding light on disease, and on how it might be treated – particularly through a process called RNA interference (RNAi). This natural phenomenon, first discovered in petunia plants in the early 1990s, is thought to have evolved as a defence against viruses. It has become one of the most exciting frontiers in medicine, in such short order that two of its pioneers, Andrew Fire and Craig Mello, won the 2006 Nobel Prize for Medicine, just eight years after their key work was published.

RNAi relies on double-stranded RNA molecules called short interfering RNAs (siRNAs), each about 21 units in length. Working with nematode worms, Fire and Mello established that when siRNAs with a particular sequence are injected into a cell, they interfere with the activity of genes that generate the same sequence in messenger RNA, so that lower quantities of protein are produced.

What happens is that once inside a cell, siRNAs are unzipped into single strands, which then bind to pieces of mRNA that match their sequence. When mRNAs are tagged in this way, they are torn up by cellular enzymes. The protein-making instructions that they carry are destroyed, and protein manufacture is impaired.

The technique's medical potential lies in its ability to target particular genes and their protein products with great precision. The 21-letter code of siRNAs can be written to match a specific set of mRNA instructions, so that production of one protein is inhibited without affecting the synthesis of others. RNAi can therefore be used to switch off rogue genes, of the sort that drive cancer or other disorders, without messing up the chemistry of healthy cells. It also allows scientists to manipulate the activity of genes in the laboratory, to establish how they work.

No RNAi drug is yet available, but several are in the advanced stages of development. Clinical trials are assessing treatments for age-related macular degeneration, a common form of blindness, which work by targeting a growth factor that is over-expressed in the eyes. Another study has shown that siRNAs can make breast tumour cells 10,000 times more sensitive to chemotherapy, by silencing genes that confer resistance to the drug Taxol. Scientists also hope to exploit the technique against HIV, to knock out a gene the virus needs to reproduce.

As science reveals more about how genes and the proteins they produce affect the course of disease, RNAi is likely to become ever more important to medicine. It is promising to provide something that clinical geneticists have long craved – a precision instrument with which to switch off genes that cause disease.

An RNAi contraceptive?

One exciting future application of RNAi could be in a new type of contraceptive pill that does not rely on hormones. Zev Williams, of Brigham and Women's Hospital in Boston, has shown that the technique can be used to silence a gene called ZP3, which is active in eggs before ovulation. When ZP3 is switched off, the egg forms without its outer membrane, to which sperm must bind for fertilization to occur.

As ZP3 is expressed only in growing eggs, the technique should be reversible: undeveloped eggs would be untouched, and could be ovulated normally once a woman has stopped taking the drug. It should also be free from side-effects, as ZP3 is not active in any other types of tissue.

the condensed idea
RNA regulates the genome

49 Artificial life

Craig Venter: 'I want to take us far from shore into unknown waters, to a new phase of evolution, to the day when one DNA-based species can sit down at a computer to design another. I plan to show that we understand the software of life by creating new artificial life.'

Mycoplasma genitalium is a bacterium that makes its home in the human urethra, where it sometimes causes a mild sexually transmitted infection. Until recently, its only distinction was to have the smallest genome of any free-living bacterium – but no longer. It has become the template for the first attempt to create artificial life.

The prospect of breathing life into inanimate matter has long intrigued humanity, as the enduring popularity of Mary Shelley's *Frankenstein* shows. A project led by Craig Venter, the maverick who led the private effort to sequence the human genome, is now promising to move it beyond science fiction and into reality.

Since 1999, he has been studying M. *genitalium* with a view to identifying the qualities of what he calls the 'minimal genome' – the smallest set of genes that is sufficient to sustain life. And now that he has reached an answer – the bacterium can survive with just 381 of the 485 genes it possesses in nature – he is seeking to synthesize such an organism in the laboratory, with a man-made genetic code. If it works, life will have been made out of chemicals from a bottle. As one of Venter's critics has put it, 'God has competition.'

Creating Synthia Venter calls the organism he plans to make *Mycoplasma labatorium*, but the ETC Group, an anti-biotech organization, has come up with a catchier name: Synthia. She would not be quite the first synthetic organism: Eckard Wimmer, of Stony Brook University, has

timeline

1999
Craig Venter (b.1946) launches minimal genome project

2002
Researchers piece together polio virus from scratch

Bioerror

Though mousepox is a relative of the smallpox virus, it does not normally make the rodents that contract it especially ill. That changed, however, when scientists at the Australian National University introduced a small genetic change to the virus in 2001. Though they had no intention of making the pathogen more virulent – they were investigating a contraceptive vaccine – the genetic modification had catastrophic effects. All the animals that were infected in the experiment died – victims not of bioterror, but of bioerror.

Critics of synthetic biology argue that if such an accident can happen when just one gene is altered in a microorganism, the potential for unintended disaster could be vast when whole genomes are being designed from scratch. Supporters say that such organisms would not be released from the laboratory until they had been proven safe, and that any that escaped accidentally would not survive in the wild.

assembled the genome of the polio virus, and Venter's team has recreated a different virus, Phi-X174, from scratch. Viruses, however, are relatively easy pickings for synthetic biology. Their genomes are tiny, and as they need to hijack host cells to reproduce, they are not normally considered to be properly alive.

Though Synthia will have a genetic code 18 times longer than any virus, her genome will also be animated, in part, by another form of life. While her DNA will be stitched together in the laboratory, scientists cannot yet reproduce the complicated cellular machinery that exists outside the nucleus, so the artificial genome will have to be transplanted into the shell of a similar bacterium. In 2007, Venter showed that such a transplant was possible, by moving the genome from one *Mycoplasma* bacterium into a close relative. This silenced the host's genome, essentially transforming one species into another. Use the same procedure to transplant a synthetic genome, and an artificial organism would enter the world.

2003
Venter's team reconstructs whole genome of phage virus Phi-X174 from scratch

2007
Venter's team creates synthetic chromosome, and transplants chromosome from one organism into another

Voyage of the Sorcerer II

Aside from genetics, Craig Venter's great passion is sailing, and he has recently combined the two in an innovative project that, he hopes, will support his efforts to make artificial life. In 2007, he published the first results of his Global Ocean Sampling Expedition, in which his yacht, the Sorcerer II, collected millions of microorganisms from the sea while sailing along the coasts of the Americas.

His goal is to identify new species, some of which might contain new genes that allow them to generate hydrogen, or store carbon dioxide. These could then be engineered into new synthetic life forms, with which to address global warming and produce green fuels.

The next stage – the construction of such a synthetic genome – has also been accomplished. Venter has rebuilt the single circular chromosome of M. *genitalium*, which contains almost 583,000 base pairs, from bottled DNA. The bacterium's code was first split into 101 chunks or 'cassettes' of 5,000 to 7,000 nucleotides, then these components were ordered from companies that manufacture short DNA sequences and assembled. The result was identical to the wild bacterium's genome, except in one important respect. A single gene, which allows wild-type M. *genitalium* to infect mammalian cells, was knocked out as insurance against accidents.

All that remains, at the time of writing, is for this synthetic chromosome to be successfully transplanted into a bacterial casing. The resulting organism will have natural hardware, but the genetic software it 'boots up' will be laboratory-made.

Use and abuse Venter's experiments with synthetic biology have two main goals. The first is intellectual – to understand a little more about the mystery of what separates living from non-living things. The second is practical – he sees it as a tool with which to manufacture organisms that benefit humankind.

Hydrogen, which some bacteria make naturally, is widely considered to be one of the energy sources of the future, as when burned it emits only water as waste. Venter plans to use synthetic biology to design microorganisms that make this clean fuel efficiently. His work is partially supported by the US Department of Energy. Other prospects include building organisms that consume and thus clean up toxic waste that is not normally biodegradable by natural bacteria, or which absorb carbon dioxide from the atmosphere to counter climate change.

Genetic engineering of existing bacteria could help to address this technological challenge, but is limited by the natural properties of the microorganisms that can be engineered. If it works, synthetic biology would open a much more powerful approach, allowing genomes to be designed from scratch with a particular purpose in mind.

Any technology, however, can be employed for both good and evil ends. Quite apart from raising moral objections from those who think it wrong to meddle with life, synthetic biology has stirred great concern about how it might be abused. As Hamilton Smith, Venter's collaborator, admitted when the team rebuilt the Phi-X174 virus: 'We could make the smallpox genome.' A deadly pathogen that has been eradicated in the wild could potentially be revived by bioterrorists or rogue states.

Just as troubling is the prospect of 'bioerror' – the accidental creation of a germ of new virulence or infectivity, against which our unhabituated bodies would have no defence. Some biologists have argued for a temporary moratorium on this research until implications and safety protocols can be worked out, as the Asilomar conference did for recombinant DNA in the 1970s (see Chapter 10).

Some of these fears are misplaced, at least for now. Venter interrupted his work for 18 months while it was reviewed by an independent ethics panel, and the microorganisms that Venter is designing will be so weak that they could not survive outside the laboratory. Genetic engineering of bacteria has also been happening for more than three decades, without a single notable accident. But as science moves forward, this technology is certain to create challenges and threats, as well as opportunities. There is a good case for proceeding with caution.

the condensed idea
Artificial life is on the way

50 No such thing as normal

Robert Plomin: 'It is not usually a case of having a disorder or not – there is quantitative variation and there is a continuum.'

The discoveries that have been described in this book should make it plain that our genomes affect virtually every aspect of human life and experience. At the most basic species level, DNA and RNA explain why we are people and not chimpanzees, mice or fruit flies. Evolved genetic change has furnished *Homo sapiens* with capacities such as language and contemplative thought, even if we so far understand only a little about the sequences that are responsible.

Within the human species, genetic variation also underlies much of its diversity. It makes an intimate contribution to every person's individuality. Dozens of genetic variants influence common medical conditions such as cancers and heart disease. Others sculpt our bodies, contributing to height, weight and looks, and still more help to shape our minds. Though science has so far located few of the alleles that mould intelligence, behaviour and personality, there is little doubt that they exist. Everybody is to some extent formed by the genetic code that he or she inherits.

Unless you are an identical twin, your genome is unique. The actual variations in DNA spelling, copy number, RNA and epigenetic programming that combine to generate these one-off profiles, however, are anything but rare. The vast majority are common – it is the idiosyncratic configurations in which they are thrown together, and the environments in which they exist, that make us all different.

1953
Identification of structure of DNA

1990s
Identification of first rare disease mutations by linkage analysis

What this means is that very little human genetic variation is abnormal. Or to look at things another way, most of it is. While there are conserved genetic sequences without which a healthy life is impossible, much of our DNA is not standard issue, from which it is unusual to deviate. We are all genetic deviants, and there is no such thing as normal.

The genetic continuum Only a few diseases, most of them rare, are a result of genetic predestination, the inevitable manifestations of single, abnormal mutations. Most others, along with traits such as intelligence, are instead influenced by hundreds of common variants. Each is carried by millions if not billions of people, and acts in concert with environmental cues and other genes and genetic processes.

A recently discovered allele that raises the chances of developing multiple sclerosis, for example, is present in about 90 per cent of Caucasians. Two-thirds of us have at least one copy of the 'fat' variant of the FTO gene. These can only be standard variants, which do not, by themselves, do anything particularly nasty. Indeed, some probably confer small benefits as

The autistic spectrum

A prime example of a genetically influenced condition that occurs on a continuum is autism, which affects people so differently that it is not usually considered a single phenomenon, but a collection of autistic spectrum disorders.

At one end of the spectrum is a highly disabling condition, characterized by social impairment, communication difficulties and non-social problems such as repetitive and restricted behaviour. At the other, people with Asperger's syndrome live perfectly independent lives, and most would consider themselves to be simply different from others, if somewhat eccentric.

Some people meet the diagnostic criteria for only one of this triad of symptoms. Many others who receive no diagnosis at all show mild versions of one or more of these traits. Autism, and the genes that affect it, seem to be an aspect of normal human variation.

2001
Completion of first drafts of human genome reveals surprisingly few genes

2006
Discovery of widespread copy number variation

2007
Genome-wide association studies identify common genetic variants linked to disease

well as small risks, such as those that seem to protect against diabetes while slightly raising predisposition to cancer.

What these do is place each of us on a continuum of normal human variation. Genetics is seldom an all-or-nothing affair, in which you inherit a particular characteristic or disease purely because you inherit a particular gene. It is usually a sliding scale, in which different genetic combinations combine with environmental factors to produce different quantitative effects.

Reading and mathematical abilities are cases in point. Both are affected by genetic variation, but as American professor Robert Plomin's research has shown, there are no genes with major effects on dyslexia or dyscalculia, let alone 'genes for' these learning difficulties. Rather, it is likely that dozens of genes with tiny effects influence literacy and numeracy skills. Genetic profiles contribute to a spectrum of ability – a few people are exceptionally gifted, most are competent to a greater or lesser extent, and a few have disabling impairments.

'One of the messages here is that the abnormal is normal,' Plomin says. 'That which we call disorder is merely one quantitative end of the normal distribution of genetic and environmental effects.' We shouldn't think that some people have genetic problems, while the rest of us are normal and fine. We all have genetic abnormalities – it is just that these differ from person to person.

Environmental engineering The fact that most genetic effects on health and behaviour are small and interactive has a further implication. It is that trying to treat and prevent disease by altering the genome is often likely to be fruitless. The experience of gene therapy illustrates the difficulty of correcting even major, single-gene mutations. Once we start to consider conditions such as diabetes, in which dozens of normal variants raise risk by a few per cent each, the idea of modifying them all starts to look fanciful. Even were it achievable, it might not be desirable: as common variants, they probably have beneficial functions as well. Careless manipulation could inflict dangerous collateral damage.

That does not mean, however, that genetic discoveries are useless – quite the reverse. For in most cases, these genes do not act alone, but hand-in-hand with the environment. A better understanding of one variable will shed light on the parallel influence of the other, and if the first is hard to

Obesity

The traditional excuse for putting on weight has always been 'big bones'. The discovery of the FTO gene has created another: 'big genes'. People who inherit one version of the gene rather than another are 70 per cent more likely to be obese. The one in six people who have the most vulnerable genotype weigh an average of 3kg more than those with the lowest risk, and also have 15 per cent more body fat.

FTO, however, is not a 'fat gene' that will inevitably make you obese. It is one of many that affect a continuum of obesity risk, to which diet and exercise matter too. If you have a 'thin' genetic profile, but fill up on pizza and hamburgers, you are still going to put on weight. And plenty of people with 'fat' variants are slim because they eat well and take regular exercise.

control, the second is often more malleable. By investigating how genetics affects bodies and minds, science can gain powerful insights into which non-genetic factors matter, and how these might be changed.

Women genetically prone to breast cancer might be referred for frequent screening, and people predisposed to diabetes could avoid diets that aggravate their genetic risk. There will also be opportunities for targeted interventions designed around knowledge of individuals' genes. Autism or dyslexia might be divided into genetic subtypes, and education programmes developed to suit one or another. Drugs, too, could be designed to change the biochemical environment in which risk genes operate. All these approaches, which might loosely be described as 'environmental engineering', are often going to trump genetic engineering and gene therapy in the genomic age.

The normality of most of the genes that affect common diseases does not imply that nothing can be done about them. Their identification will allow science to investigate these conditions from a position of knowledge, and thus from a position of power.

the condensed idea
Genetic variation is a continuum

Glossary

Allele Alternative variant of a gene. Individuals will generally have two alleles of each gene, which may vary

Amino acids Molecules from which proteins are built. All life uses 20 different amino acids, the instructions of which are carried by codons or triplets of DNA and RNA

Autosome Non-sex chromosome, which always has a matching pair. Humans have 22 pairs of autosomes

Base pair A pair of complementary bases or nucleotides (A and T or C and G)

Behavioural genetics The study of genetic factors that affect non-medical traits, such as intelligence or personality

Caenorhabditis elegans Species of microscopic nematode worm, commonly used in genetic research

Centromere Central point of a chromosome, at which its short and long arms join

Chromatin The complex of DNA and histone proteins of which chromosomes are composed. Chromatin can be modified to alter gene expression

Chromosome Strand of DNA that carries genes and other genetic information. Humans have 46, consisting of 22 pairs of autosomes, and 2 sex chromosomes

Clone 1. A piece of DNA that has been reproduced in a bacterium, for study or sequencing
2. An organism created by replicating the nuclear DNA of an adult organism, usually achieved by somatic cell nuclear transfer

Codon (triplet) A string of three DNA or RNA bases that codes for production of an amino acid

Copy number variation Duplications or deletions of DNA sequences, which can differ between individuals

DNA Deoxyribonucleic acid, the molecule that holds the genetic instructions for most forms of life. Its structure is a double helix

Dominant Allele that is always expressed, even when it differs from the other allele, as in heterozygotes

Double helix *see* DNA

Drosophila melanogaster Species of fruit fly, commonly used in genetic research

Enzyme Specialized form of protein, which catalyses chemical reactions in cells

Epigenetics The phenomenon by which chemical modifications to DNA and chromatin alter gene expression, without altering the genetic code itself

Exons The units within genes that contain protein-coding information. They are broken up by introns

Gamete Reproductive cells, which contain only half the usual complement of chromosomes. In humans, these are sperm and eggs, which each contain 23 unpaired chromosomes

Gene The fundamental unit of inheritance. Usually taken to mean a section of DNA that contains the code for a protein, but the definition is being widened to include DNA that carries other genetic instructions

Gene expression Process by which genes are switched on or off in cells

Genetic drift An evolutionary process by which genes can become more or less common without natural selection

Genetic fingerprint A repetitive section of DNA which can identify an individual reasonably reliably. Used in forensic science

Gene therapy Medical technique for correcting faulty genes that cause disease, usually involving a genetically modified virus

Genome The full genetic code of an organism

Genome-wide association Gene-hunting technique capable of finding genes with small effects on diseases or other phenotypes

Genotype An individual's genetic profile. This can refer to one or many different alleles

Germ cell Adult cell which is the precursor of gametes

Haplotype A sub-section of a chromosome that tends to remain intact during recombination. Haplotype blocks are responsible for genetic linkage

HapMap A map of common haplotypes in four ethnic groups, which is now used extensively to guide genetic research

Heritability A measure of how much inheritance contributes to the variability of a phenotype, usually expressed as a decimal or a percentage

Heterozygote (Heterozygous adj.) An individual with two different alleles of a particular gene or DNA sequence

Homozygote (Homozygous adj.) An individual with two identical alleles of a particular gene or DNA sequence

Imprinted gene A gene marked so it is expressed according to maternal or paternal origin

In-vitro fertilization (IVF) Assisted reproduction procedure, by which eggs are fertilized by sperm in the laboratory, and resulting embryos are then transferred to the womb

Introns Non-coding passages of DNA that break up the coding regions, or exons, of genes

Junk DNA DNA that does not code for protein. Much of it, however, is transcribed into RNA and regulates gene expression

Linkage Phenomenon by which certain alleles tend to be inherited together, as they lie close together on a chromosome

Meiosis Cell division process by which germ cells create gametes. Cells produced by meiosis have only one set of chromosomes instead of the usual two. Recombination occurs during meiosis

Mendelian trait A characteristic that is passed on by simple dominant or recessive genes

Messenger RNA (mRNA) The adaptor molecule into which protein-coding DNA is transcribed, which carries the instructions for making a protein

Methylation Process by which DNA is chemically modified, often associated with gene silencing. Important to epigenetics and imprinting

Mitochondria Cellular structures outside the nucleus that generate energy, which contain DNA. Mitochondria are always inherited from the mother, and mitochondrial DNA is useful for tracing maternal ancestry

Mitosis Normal process of cell division, by which a cell copies its genetic material and splits. The resulting daughter cells carry the same DNA as their parents, save for any random mutations

Mutation Process by which the DNA sequence is altered, by substitution of one base for another. It can occur randomly, as a result of copying errors, or through damage by radiation or chemicals

Natural selection Main process of evolution, by which organisms that acquire beneficial mutations have greater reproductive success

Nucleotide (base) A DNA or RNA 'letter', in which the genetic code is spelled. DNA's nucleotides are adenine (A), cytosine (C), guanine (G) and thymine (T). In RNA, uracil (U) takes the place of thymine

Nucleus Cellular structure containing the chromosomes and most of an organism's DNA. Organisms with nuclei are known as eukaryotes

Oncogene A gene that when mutated can promote unchecked cell division and cancer

Phage (bacteriophage) A kind of virus that infects bacteria, often used in genetic research

Pharmacogenomics The field of prescribing drugs according to a patient's genetic profile

Pharming Colloquial term for using genetically modified animals to make industrial or medical products

Phenotype An observed characteristic of an organism, which can be influenced by either inheritance or the environment

Plasmid A circular strand of bacterial DNA that exists outside the chromosome. Often used in genetic engineering

Pre-implantation genetic diagnosis (PGD) Technique by which a single cell is removed from IVF embryos and checked for the presence of particular genes or chromosomes. Often used to screen for inherited disease

Protein Large organic compound made from a long chain of amino acids. Many proteins are enzymes that catalyse chemical reactions in cells. Others are structural, such as collagen

Recessive Allele that is expressed only when two copies are present, in homozygotes

Recombinant DNA An artificial strand of DNA created by genetic engineering. Often used to produce drugs from bacteria

Recombination (crossing-over) Process that occurs during meiosis, by which chromosomes exchange chunks of genetic material

Regulatory region DNA sequence that alters the activity of other DNA sequences

Replication Process by which the DNA double helix unzips to copy itself

Restriction enzyme Chemical that cuts DNA when a particular sequence appears, commonly used as 'molecular scissors' in genetic engineering

Ribosome Cellular structure made of RNA and protein that uses messenger RNA instructions to produce protein

RNA Ribonucleic acid, a chemical cousin of DNA, usually single-stranded, which carries genetic messages inside cells

RNA interference Process by which small RNA molecules can silence the production of particular proteins

Sequencing Method for reading the code of individual genes or the genomes of entire species

Sex chromosome Chromosomes that determine an organism's sex, such as the X and Y chromosomes in humans. The genotype XX is female, and XY is male

Single nucleotide polymorphism (SNP) Point at which the genetic code commonly differs by one base between individuals of the same species. Standard form of genetic variation

Somatic cell Specialized adult cell that divides by mitosis. Includes all specialized cell types except germ cells, gametes and undifferentiated stem cells

Somatic cell nuclear transfer Cloning procedure by which the nucleus of a somatic cell is transferred into an egg from which the nucleus has been removed

Splicing Process by which introns are removed from messenger RNA prior to translation into protein

SRY The sex-determining region Y gene, which establishes male sex

Stem cell Undifferentiated cell with the potential to make multiple tissue types. The most versatile are embryonic stem cells, which can make every tissue type

Telomere Repetitive DNA structure found at the ends of chromosomes, which protects them against damage during replication and cell division

Transcription Process by which DNA is copied into RNA to make proteins and regulate gene expression

Translation Process by which messenger RNA is used to manufacture protein

Tumour suppressor Type of gene that identifies potentially cancerous mutations and induces cell suicide. Often mutated in tumours

Twin study Common tool for genetic research, by which identical twins who share all their DNA are compared with fraternal twins who share only half of it

Index

Quercus Publishing Plc
21 Bloomsbury Square
London WC1A 2NS

First published in 2008

Copyright © Mark Henderson 2008

Mark Henderson has asserted his right to be identified as the author of this Work.

All rights reserved. No part of this publication may be reproduced, stored in a retrieval system, or transmitted in any form or by any means, electronic, mechanical, photocopying, recording, or otherwise, without the prior permission in writing of the copyright owner and publisher.

Quercus Publishing Plc hereby exclude all liability to the extent permitted by law for any errors or omissions in this book and for any loss, damage or expense (whether direct or indirect) suffered by a third party relying on any information contained in this book.

A catalogue record of this book is available from the British Library

Cloth case edition ISBN-13: 978 1 84724 671 4
Printed case edition ISBN-13: 978 1 84724 670 7

Printed and bound in China

10 9 8 7 6 5 4 3 2 1

Prepared by *specialist* publishing services ltd, Montgomery